Edited by
I. Gohberg

Birkhäuser
Boston · Basel · Stuttgart

S. G. Krein

Linear Equations in Banach Spaces

Translated from the Russian
by A. Iacob

1982

Birkhäuser
Boston • Basel • Stuttgart

Author:

S. G. Krein
Department of Mathematics
Voronezh University
Voronezh, USSR

Library of Congress Cataloging in Publication Data

Kreĭn, S. G. (Selim Grigorevich), 1917-
Linear equations in Banach spaces.

Translation of: Lineinye uravneniia v banakhovom prostranstve.
Bibliography: p.
1. Operator equations. 2. Differential equations, Linear. 3. Banach spaces. 4. Integral equations.
I. Iacob, A. II. Title.
QA329.2.K7413 1982 515.7'246 82-17736
ISBN 3-7643-3101-1

CIP- Kurztitelaufnahme der Deutschen Bibliothek

Krejn, Selim G.:
Linear equations in banach spaces / S. G. Krein.
Transl. from the Russian by A. Iacob. - Boston;
Basel ; Stuttgart : Birkhäuser, 1982.
(Gohberg-textbook-series)
ISBN 3-7643-3101-1

ISBN 3-7643-3101-1

Printed in USA

EDITORIAL INTRODUCTION

This slim volume contains a general theory of linear equations in Banach spaces with applications to differential and integral equations. It contains a detailed analysis of the uniqueness and correctness problem, of over-determined and under-determined equations, connections between the equation and its conjugate. Different transformations of linear equations are considered. Some sections deal with problems of stability and apriori estimates. Different classes of equations, such as Fredholm equations, Noether equations, and others are considered. Applications are given to Fredholm integral equations, Volterra integral equations, boundary problems for linear differential equations of higher orders, and boundary problems for elliptic partial differential equations.

To understand the book it is enough to have a limited knowledge of the basics of functional analysis. An appendix at the end of the book contains the statements of the theorems of functional analysis which are used in the main text, and this will be found useful to the reader.

This book differs from others in its conciseness. It keeps its aim clearly in sight and does not stray from that aim; it is systematic and not difficult to read. It is uncanny how much material the author was able to include in this slim volume.

The author is an outstanding expert in differential equations, the theory of linear operators and applied mathematics. This fact influences the choice of material, the presentation and terminology. One clearly feels the applied motives in the book.

The book may be recommended for senior undergraduate students and graduates in mathematics and related fields. In my opinion it is of interest to anybody dealing with linear equations.

I would like to thank A.Iacob for the excellent job he did in translating this volume.

I.Gohberg

TABLE OF CONTENTS

PREFACE

The basic material appearing in this book represents the substance of a special series of lectures given by the author at Voronež University in 1968/69, and, in part, at Dagestan University in 1970. In 1968 these lecture notes appeared in mimeographed form, at Voronež State University. It seems to us that this course may be given as soon as three of the basic principles of functional analysis and their main corollaries have been discussed, and that it may be presented as an important application of these principles (all the background material concerning functional analysis and required for this book can be found in the appendix to the text). On the other hand, these lectures could form the theoretical basis for other courses devoted to the study of one or the other class of integral or differential equations. In fact, some classes of equations of this particular type are considered in the last two sections of the book, in order to illustrate the results of the basic text. Naturally, however, many fundamental facts from the theory of integral or differential equations are given without proofs.

Much of the material discussed here is contained in books on functional analysis, in various places, and under various names. Some results are taken from papers on operator theory and on the theory of partial differential equations. We felt that it would be useful to gather together these results and to unite the whole subject into a theory of linear equations in Banach spaces with generally unbounded operators. One should remark that the monograph by D. Przeworska--Rolewicz and S. Rolewicz [13], which the author had the opportunity to consult, was apparently written with the same purpose in mind.

Here we have paid special attention both to the relations between the properties of the given equation and the properties of its adjoint equation, and to the theory of Noetherian and Fredholm equations.

The author would like to express his deep gratitude to the members of his audience T.Ja. Azizov, I.S. Gudovič, and I.Ja. Šneiberg for the indispensable help they rendered during the preparation of this book.

S. Krein

INTRODUCTION

Recall that any system of n linear equations with n unknowns can be considered to be a single equation of the form

$$Ax = y \tag{1}$$

in the n-dimensional Euclidean space E_n, where y is the given vector of right-hand sides, x is the unknown vector of solutions, and A is the linear transformation defined by the matrix of the system. In this setting, we have the following basic facts from linear algebra.

1. Equation (1) is solvable for any right-hand side if and only if the corresponding homogeneous equation

$$Ax = \theta$$

has only the trivial solution $x = \theta$. [From now on θ shall denote the zero element of the space under consideration.]

2. Equation (1) is solvable for any right-hand side if and only if the adjoint equation

$$A^*g = f \tag{2}$$

(where the matrix of the linear transformation A^* is the complex conjugate and transpose of the matrix of A) is solvable for any right-hand side.

3. The equations $Ax = \theta$ and $A^*g = \theta$ have the same number of linearly independent solutions.

4. If equation (1) is not solvable for all right-hand sides, then the set of right-hand sides y for which (1) is solvable is a subspace of the space E_n, namely the orthogonal complement of the subspace of all solutions to the homogeneous equation $A^*g = \theta$.

If we consider m linear equations with n unknowns and $n \neq m$, then the situation is slightly more complicated. Namely, if we write this system in vector form as in (1), then the operator A generated by the matrix of the system, must be considered to be an operator from the n-dimensional space E_n into the m-dimensional space E_m.

The operator A^* in the adjoint equation (2) takes E_m into E_n. If, for example, $n > m$, then the homogeneous equation $Ax = \theta$ always has a nontrivial solution, regardless of the solvability of the system (1). Therefore, Property 1 is not valid any more, and the connection between the solvability of (1) and the solvability of (2) is also destroyed. Properties 1 and 2 can be replaced by

1°. Equation (1) is solvable for any right-hand side if and only if the homogeneous equation $A^*g = \theta$ has only the trivial solution.

2°. Equation $A^*g = f$ is solvable for any right-hand side if and only if the homogeneous equation $Ax = \theta$ has only the trivial solution.

There is also an analogue of Property 3. If the matrix A has rank r, then the homogeneous system $Ax = \theta$ has $n - r$ linearly independent solutions, while the homogeneous system $A^*g = \theta$ has $m - r$ such solutions. We see that these numbers are not equal, but their difference, $n - m$, is independent of the operator A.

3°. For all equations (1), the difference between the maximal number of linearly independent solutions of the homogeneous equation and the maximal number of linearly independent solutions of the homogeneous adjoint equation is the same and equals $n - m$.

The number $n - m$ is called the *index* of the system and can be positive as well as negative.

Property 4 remains true (4 = 4°).

When linear equations in an infinite dimensional space are considered, the variety of possible cases is greater. However, some similarities to the finite dimensional case remain.

L I N E A R   E Q U A T I O N S   I N   B A N A C H   S P A C E

by

S. G. KREIN

Translated from the Russian by A. Iacob

§1. LINEAR EQUATIONS. BASIC NOTIONS

Let E be a Banach space and assume that a linear operator A is defined on some linear manifold $\mathcal{D}(A)$ in E and that A takes $\mathcal{D}(A)$ into a Banach space F. The set $\mathcal{D}(A)$ is called the *domain of definition* or the *domain* of the operator A.

Consider the equation

$$Ax = y, \tag{A}$$

where y is a given element of F, and x is the unknown element in $\mathcal{D}(A)$. The collection of all $y \in F$ such that equation (A) is solvable is a linear manifold in F, called the *range* $R(A)$ of the operator A. The collection of all the solutions of the corresponding homogeneous equation $Ax = \theta$ is a linear manifold in E, called the *null-space* or the *kernel* $N(A)$ of the operator A. [Sometimes the range of the operator A is denoted by $\operatorname{Im} A$ (image of A), and its kernel — by $\operatorname{Ker} A$.]

Equation (A) is said to be *uniquely solvable on* $R(A)$ if the homogeneous equation $Ax = \theta$ has only the null solution, i.e., if $N(A) = \theta$. In this case, for each $y \in R(A)$, there is only one solution of the equation $Ax = y$, and so the operator A has an inverse A^{-1} on $R(A)$: $A^{-1}y = x$ $(y \in R(A))$.

Equation (A) is said to be *correctly solvable on* $R(A)$ if the inequality $\|x\|_E \leq k\|Ax\|_F$ holds for all $x \in \mathcal{D}(A)$, where $k > 0$ and does not depend upon x. Correct solvability implies unique solvability. In the finite dimensional case, the two conditions are equivalent.

If equation (A) is correctly solvable, then the operator A has a bounded inverse on $R(A)$.

Equation (A) is *normally solvable* if $R(A)$ is a (closed !) subspace of F: $R(A) = \overline{R(A)}$. [Translator's note. From now on, one adheres to the convention: a *subspace* is a *closed* linear manifold.]

Equation (A) is *densely solvable* if $R(A)$ is dense in F: $\overline{R(A)} = F$. Finally, equation (A) is *everywhere solvable* if $R(A) = F$.

Assume that the kernel $N(A)$ of the operator A is closed. Then one can take the quotient space $E/N(A)$ whose elements are cosets of the elements $x \in E$ relative to the subspace $N(A)$: $X = \{x + z\}$ $(z \in (A))$. The quotient space is a Banach space relative to the norm

$$\|X\| = \inf_{z \in N(A)} \|x - z\| .$$

The following holds true: *if the sequence* X_n *is fundamental, then there is a sequence* n_k *and a sequence* $x_{n_k} \in X_{n_k}$ *such that* $\{x_{n_k}\}$ *converges.*

To prove this, choose a sequence n_k such that $\|X_{n_k} - X_{n_{k-1}}\| < < 1/2^k$. Next, find an element $z_k \in X_{n_k} - X_{n_{k-1}}$ such that $\|z_k\| < 1/2^k$. Then the series $x = x_{n_0} + z_1 + z_2 + \dots$ converges, and, letting $x_{n_k} = x_{n_0} + z_1 + z_2 + \dots + z_k$, one has

$$x_{n_k} \longrightarrow x \quad \text{and} \quad x_{n_k} \in X_{n_0} + \textstyle\sum_{r=1}^{k}(X_{n_r} - X_{n_{r-1}}) = X_{n_k}.$$

Using the above statement, we can easily deduce that the space $E/N(A)$ is complete. Indeed, in the previous notations, the class X containing the constructed element x is the limit of the sequence X_n in $E/N(A)$.

PROPOSITION 1.1. *The space* $E/N(A)$ *is complete. The convergence of the sequence* $X_n \longrightarrow X$ *is equivalent to the existence of both a sequence* $x_n \in X_n$ *and an element* $x \in X$ *such that* $x_n \longrightarrow x$.

To prove the second statement, observe that $x_n - x \longrightarrow 0$ implies $\|X - X_n\| \leq \|x - x_n\| \longrightarrow 0$. Conversely, if $\|X - X_n\| \longrightarrow 0$, then there is a sequence $u_n \in X - X_n$ such that $\|u_n\| \longrightarrow 0$. Now $u_n = x^{(n)} - x_n^{(n)}$, where $x^{(n)} \in X$ and $x_n^{(n)} \in X_n$. Let $x \in X$ be fixed. Then

$$u_n = x + z^{(n)} - x_n^{(n)} = x - (x_n^{(n)} - z^{(n)}),$$

where $z^{(n)} \in N(A)$. If we denote $x_n = x_n^{(n)} - z^{(n)} \in X_n$, we see that $x_n \longrightarrow x$, which completes the proof of the proposition.

In a natural way, the operator A induces a linear operator $\tilde{A}$ defined on $\mathcal{D}(A)/N(A)$ by the formula: $\tilde{A}X = Ax$ $(x \in \mathcal{D}(A))$. The kernel of $\tilde{A}$ reduces to zero, i.e., the *factored* or *quotient* equation

$$\tilde{A}X = y \qquad (\tilde{A})$$

is uniquely solvable.

Let us see what the correct solvability of the equation $(\tilde{A})$, i.e., the assumption $\|X\| \leq k_1\|\tilde{A}X\|$, means to the initial equation (A). By the definition of the norm in $E/N(A)$, given any $\varepsilon > 0$ one can find an element $x \in X$ such that $\|X\| \leq \|x\| \leq \|X\| + \varepsilon$.

Choose $\varepsilon = k_1\|\tilde{A}X\|$. Then

$$\|x\| \leq 2k_1\|\tilde{A}X\| = k\|Ax\| \qquad (k = 2k_1).$$

Conversely, if there is $k > 0$ such that for any $y \in R(A)$ there is at least one element $x \in \mathcal{D}(A)$ satisfying $\|x\| \leq k\|Ax\|$, then $\|X\| \leq \|x\| \leq k\|Ax\| = k\|\tilde{A}X\|$.

Therefore, *the factored equation* $(\tilde{A})$ *is correctly solvable if and only if there exists* $k > 0$ *such that, given any* $y \in R(A)$, *there is at least one solution* x *of* (A) *satisfying* $\|x\| \leq k\|Ax\| = k\|y\|$.

OBSERVATION 1.2. Assume that the last inequality holds true. If the sequence Ax_n converges, then so does the sequence $\tilde{A}X_n$, which in virtue of the correct solvability of $(\tilde{A})$, implies that the sequence X_n converges. Now apply Proposition 1.1 to show that there exists a convergent sequence $\bar{x}_n$ with $\bar{x}_n \in X_n$, and so we have $A\bar{x}_n = \tilde{A}X_n$.

§ 2. EQUATIONS WITH A CLOSED OPERATOR

A linear operator A is called *closed* if whenever $x_n \longrightarrow x$ and $Ax_n \longrightarrow y$ we have $x \in \mathcal{D}(A)$ and $Ax = y$.

Each bounded operator, defined on the whole space, is closed, but a closed operator is not necessarily bounded. The study of closed operators is often reduced to the study of bounded operators using the following artifice: a new norm (the *graph norm*),

$$\|x\|_{E_A} = \|x\|_E + \|Ax\|_F,$$

is introduced on the domain $\mathcal{D}(A)$ of A. Referring to the definition of a closed operator, we see that the space $E_A = \mathcal{D}(A)$ endowed with this new norm is complete, i.e., E_A is a Banach space. The operator A is defined on E_A and is bounded, because $\|Ax\|_F \leq \|x\|_{E_A}$.

THE BANACH THEOREM. *A closed operator which is defined on the entire space* E *is bounded.*

PROOF. Under our assumption, $E_A = E$ and thus on E there are two norms relative to which E is complete, and we have $\|x\|_E \leq \|x\|_{E_A}$. Therefore, these two norms are equivalent, i.e.,

$$\|x\|_{E_A} \leq c\|x\|_E, \text{ and so } \|Ax\|_F \leq c\|x\|_E.$$

The theorem is proven.

If a closed operator A has an inverse on $R(A)$ $(N(A) = \Theta)$, then it is easy to check that the inverse A^{-1} is also closed. This observation allows us to prove the following important fact.

THEOREM 2.1. *Let* A *be a closed operator. Then equation* (A) *is correctly solvable if and only if it is uniquely and normally solvable.*

PRROF. If equation (A) is correctly solvable, then it is uniquely solvable. Let us check that it is also normally solvable, i.e., that $R(A)$ is closed. Let $y_n = Ax_n \in R(A)$ and $y_n \longrightarrow y$. Then

$$\|x_n - x_m\|_E \leq k\|A(x_n - x_m)\|_F = k\|y_n - y_m\|_F,$$

and so the sequence x_n converges to an element x. A being a closed operator, $x \in \mathcal{D}(A)$ and $y = Ax \in \mathcal{R}(A)$.

Conversely, assume that equation (A) is uniquely and normally solvable. Then the inverse operator A^{-1} on the Banach space $\mathcal{R}(A)$ is defined and A^{-1} is closed, and hence, by Banach's theorem, bounded. This means that $\|A^{-1}y\|_E \leq k\|y\|_F$ for all $y \in \mathcal{R}(A)$, or

$$\|x\|_E \leq k\|Ax\|_F \qquad (x \in \mathcal{D}(A)).$$

In other words, (A) is correctly solvable, and the proof of the theorem is complete.

Given a closed operator A, the kernel $N(A)$ is closed and thus one can consider the factored equation $(\tilde{A})$.

PROPOSITION 2.2. *If* A *is closed, then* $\tilde{A}$ *is also closed.*

PROOF. Let $X_n \longrightarrow X$, $X_n \in \mathcal{D}(\tilde{A})$, and $\tilde{A}X_n = y_n \longrightarrow y$. By Proposition 1.1, there exists a sequence $x_n \in X_n$ which converges to an element $x \in X$. Then $Ax_n = \tilde{A}X_n \longrightarrow y$, and since A is closed, we obtain $x \in \mathcal{D}(A)$ and $y = Ax$. Therefore $X \in \mathcal{D}(\tilde{A})$ and $\tilde{A}X = Ax = y$, which proves the proposition.

Equation $(\tilde{A})$ is always uniquely solvable. According to Theorem 2.1, $(\tilde{A})$ is correctly solvable if and only if it is normally solvable. On the other hand, equations (A) and $(\tilde{A})$ are simultaneuously normally solvable or not $(\mathcal{R}(A) = \mathcal{R}(\tilde{A}))$. Apllying the criterion for the correct solvability to equation (A), we obtain the following result.

THEOREM 2.3. *Let the operator* A *be closed. Then equation* (A) *is normally solvable if and only if the following condition is satisfied: there exists a number* $k > 0$ *such that given any* $y \in \mathcal{R}(A)$, *there is* $x \in \mathcal{D}(A)$ *with* $y = Ax$ *and* $\|x\| \leq k\|Ax\| = k\|y\|$.

Theorems 2.1 and 2.3 can be written in the following schematic form.

For equations with closed operator

$$\textit{correct solv.}\ (A) \Leftrightarrow \textit{unique solv.} + \textit{normal solv.}\ (A),$$
$$\textit{correct solv.}\ (\tilde{A}) \Leftrightarrow \textit{normal solv.}\ (A).$$

We can give another sufficient condition for the normal solvability of equation (A).

THEOREM 2.4. *Consider equation* (A) *with a closed operator. If the space* F *decomposes into a direct sum* $F = \mathcal{R}(A) \oplus L$, *where* L *is a subspace of* F, *then* (A) *is normally solvable.*

PROOF. Consider again the factored equation with operator $\tilde{A}$. Endowing the domain $\mathcal{D}(\tilde{A})$ with the graph norm, we can look upon $\tilde{A}$ as a bounded operator from the new Banach space $E_{\tilde{A}}$ into F. Now define the operator $\hat{A}$ on the product $E_{\tilde{A}} \times L$ by $\hat{A}(X,v) = \tilde{A}X + v$ $(X \in E_A$ $v \in L)$. This operator is bounded and is a one-to-one mapping from $E_{\tilde{A}} \times L$ onto F.

Indeed, if $\tilde{A}X + v = \theta$, then since $\tilde{A}X \in R(A)$, we have $v = \theta$ and $\tilde{A}X = \theta$. This implies $X = \theta$. Further, any $y \in F$ can be represented as $y = Ax + v = \tilde{A}X + v$, i.e., $y \in R(\hat{A})$. It results from the open mapping theorem that the operator $\hat{A}$ has a bounded inverse $\hat{A}^{-1}$. The linear manifold of all elements of the form (X,θ) is closed in $E_{\tilde{A}} \times L$, and hence its preimage under the map $\hat{A}^{-1}$ is a closed subset of F; but this preimage equals $R(A)$.

The theorem is proven.

Equations (A) where A is a compact, but not a finite rank operator provide examples of equations with closed operators which are not normally solvable (and therefore, not correctly solvable).

Indeed, assume that the range $R(A)$ is closed, and let $Q \subset R(A)$ be an arbitrary bounded set. By Theorem 2.3, for any element $y \in Q$, one can find a preimage under A, and together these preimages, as y runs over Q, form a bounded subset M of E. But then $Q = AM$ is compact. Consequently, our subspace $R(A)$ is locally compact, and hence finite dimensional.

A nonclosed operator is said to be *closable* (or *to admit a closure*) if it can be extended to a closed operator. A linear operator A is closable if and only if the convergent sequences $x_n \longrightarrow \theta$, $Ax_n \longrightarrow y$ imply that $y = \theta$. The necessity of this condition is obvious. Now assuming that the condition is satisfied, we can define a closed extension $\bar{A}$ of the operator A by setting $\bar{A}x = y$ whenever there is a sequence $x_n \in D(A)$ such that $x_n \longrightarrow x$ and $Ax_n \longrightarrow y$. The operator $\bar{A}$ is called the *closure* of A. In this way, given a closable operator, one may consider, along with equation (A), the equation

$$\bar{A}x = y. \tag{$\bar{A}$}$$

The solutions of $(\bar{A})$ are often called *generalized solutions* of the equation (A).

EXAMPLES. Consider the space $C[0,1]$ of all function $x(t)$ continuous on $[0,1]$, and the operator $Ax = \frac{dx}{dt}$, defined on all continuously differentiable functions. Then (according to classical theorems of calculus), A is a closed operator from $C[0,1]$ into

$C[0,1]$.

The same differentiation operator, when considered as an operator from $L_2[0,1]$ into $L_2[0,1]$, is not closed. However, it is closable and its closure is called the operator of *generalized differentiation*.

Let A be the operator from $C[0,1]$ into $\mathbb{R}$ defined on the continuously differentiable functions by the formula $Ax = x'(0)$. Then A is not closable: indeed, the sequence $x_n(t) = \frac{1}{n}\sin nt$ converges to zero uniformly on $[0,1]$, while $Ax_n = x_n'(0) = 1$.

§ 3. THE ADJOINT EQUATION

In most applications, the operator A has a natural domain of definition which is a normed space. This space can be completed to a Banach space E, and thus A will have a domain $\mathcal{D}(A)$ which is dense in E. This is precisely the situation that we shall consider in this section.

Let g be a bounded linear functional on F ($g \in F^*$), and consider the linear functional $g(Ax)$ on $\mathcal{D}(A)$. This functional is not necessarily bounded. If it is, we write $g \in \mathcal{D}(A^*)$ and set, by definition,

$$(A^*g)(x) = g(Ax). \tag{3.1}$$

This formula defines the functional A^*g only on $\mathcal{D}(A)$. However, since $\mathcal{D}(A)$ is dense in E, and A^*g is bounded on $\mathcal{D}(A)$ by assumption, the functional A^*g can be extended uniquely to a bounded linear functional on E.

Consequently, we may look upon A^* as an operator mapping $\mathcal{D}(A^*) \subset F^*$ into E^*. The operator A^* is obviously linear and is called the *adjoint* of A. [Translator's note. The following equivalent terms appear frequently in the literature: *dual, conjugate,* or *transpose* of A.] We remark that its domain of definition $\mathcal{D}(A^*)$ is not necessarily dense in F^* (see an example on page 14).

The adjoint operator is always closed. Indeed, if $g_n \longrightarrow g$ and $A^*g_n \longrightarrow f$, then, for any $x \in \mathcal{D}(A)$, the first relation implies that $g_n(Ax)$ converges to $g(Ax)$, and the second relation implies convergence to $f(x)$, i.e., $g(Ax) = f(x)$ $(x \in \mathcal{D}(A))$ and $f = A^*g$.

The notions of unique, correct, dense, and everywhere solvability for the adjoint equation

$$A^*g = f \qquad (g \in \mathcal{D}(A^*) \subset F^*,\ f \in E^*) \tag{A*}$$

are defined in the same way as they were for equation (A).

Now let us see what structure the kernel of A^* has. If $g \in N(A^*)$, then

$$0 = (A^*g)(x) = g(Ax)$$

for all $x \in \mathcal{D}(A)$, i.e., the functional g is orthogonal to $\mathcal{R}(A)$. Conversely, if $g \in \mathcal{R}(A)^{\perp}$, then $g(Ax) = 0$ for all $x \in \mathcal{D}(A)$, and so $g \in \mathcal{D}(A^*)$ and $A^*g = 0$.

In other words:

The kernel of the adjoint operator is the orthogonal complement of the range of the initial operator.

A direct corollary is

THEOREM 3.1. *Equation* (A) *is densely solvable* ($\overline{\mathcal{R}(A)} = F$) *if and only if equation* (A*) *is uniquely solvable* ($N(A^*) = \theta$).

If $N(A^*) = \mathcal{R}(A)^{\perp}$, then it is not hard to see that ${}^{\perp}N(A^*) = \overline{\mathcal{R}(A)}$. Indeed, if $x \notin \overline{\mathcal{R}(A)}$, one can construct a linear functional equal to zero on the subspace $\overline{\mathcal{R}(A)}$, i.e., contained in $N(A^*)$, and different from zero at x. This shows that $x \notin {}^{\perp}N(A^*)$. At the same time, we obtain a criterion for normal solvability (which justifies this term).

THEOREM 3.2. *Equation* (A) *is normally solvable if and only if* $\mathcal{R}(A) = {}^{\perp}N(A^*)$, *i.e., if and only if equation* (A) *is solvable for precisely those right-hand sides which are orthogonal to all solutions of the homogeneous adjoint equation.*

The connection between $N(A)$ and $\mathcal{R}(A^*)$ is significantly more complicated and we analyse it now. If $x \in N(A)$ and $f \in \mathcal{R}(A^*)$, then $f(x) = (A^*g)(x) = g(Ax) = 0$, i.e., $N(A)$ and $\mathcal{R}(A^*)$ are orthogonal. We obtain

THEOREM 3.3. *Equation* (A) *is uniquely solvable on* $\mathcal{R}(A)$ *if and only if equation* (A*) *is densely solvable.*

[The following weaker condition is also sufficient: the range $\mathcal{R}(A^*)$ is weakly dense in E^* (see [7]).]

Indeed, if $Ax = \theta$, then $x \in {}^{\perp}\mathcal{R}(A^*) = {}^{\perp}\overline{\mathcal{R}(A^*)} = {}^{\perp}E^*$, and hence $x = \theta$.

The theorem is proven.

However, the sets $N(A)$ and $\mathcal{R}(A^*)$ are not necessarily the orthogonal complements of each other. Namely, the fact that a functional f is equal to zero on $N(A)$ does not imply automatically that equation (A*) is solvable for f. Therefore, we can write only the inclusion $\mathcal{R}(A^*) \subset N(A)^{\perp}$.

Conversely, the fact that $x \perp \mathcal{R}(A^*)$ does not, on the one hand, imply that $x \in \mathcal{D}(A)$; on the other hand, even if $x \in \mathcal{D}(A)$, and so $g(Ax) = 0$ for all $g \in \mathcal{D}(A^*)$, the equality $Ax = \theta$ results only if $\mathcal{D}(A^*)$ is dense in F^* in the weak topology. If operator A is bounded and defined on the entire space E, then these last two considerations are no longer valid and x belongs to $N(A)$. In this case operator A^* is also defined on the entire space F^*.

Consequently, if A is bounded, the equality

$$N(A) = {}^{\perp}R(A^*)$$

holds.

Finally, we remark that the condition $N(A) = {}^{\perp}R(A^*)$ does not necessarily imply that $N(A)^{\perp}$ equals $\overline{R(A^*)}$. In particular, the equality $R(A^*) = N(A)^{\perp}$ is not necessarily a consequence of the fact that $R(A^*)$ is closed. (One can show that under condition above $N(A)^{\perp}$ equals the closure of $R(A^*)$ in the weak topology of the space E^*). Therefore, we are justified in introducing two generally distinct notions for the adjoint equation :

1) Equation (A^*) is called *closedly solvable* if $R(A^*)$ is closed.

2) Equation (A^*) is called *normally solvable* if it is solvable for any right-hand side which is orthogonal to all the solutions of the homogeneous equation $Ax = \theta$, i.e., if $R(A^*) = N(A)^{\perp}$.

Clearly, the closed solvability of (A^*) results from its normal solvability. The discussion above shows that the converse is not true in general. Now we consider other problems.

THEOREM 3.4. *Equation* (A^*) *is everywhere solvable if and only if equation* (A) *is correctly solvable on* $R(A)$.

PROOF OF THE NECESSITY. Let (A^*) be everywhere solvable: given any $f \in E^*$, there is $g \in F^*$ such that $f = A^*g$. For $x \in D(A)$

$$|f(x)| = |(A^*g)(x)| = |g(Ax)| \leq \|g\|_{F^*}\|Ax\|_F .$$

If x is fixed, the formula $\Phi_x(f) = |f(x)|/\|Ax\|_F$ defines a convex continuous functional of $f \in E^*$. These functionals are are uniformly bounded for any fixed f:

$$\Phi_x(f) = \frac{|f(x)|}{\|Ax\|_F} \leq \|g\|_{F^*} \qquad (x \in D(A)).$$

According to the uniform boundedness principle, there exists $k > 0$ such that

$$\Phi_x(f) \leq k\|f\|_{E^*}, \quad f \in E^*$$

or

$$|f(x)| \leq k\|f\|_{E^*}\|Ax\|_F .$$

Therefore, $\|x\|_E \leq k\|Ax\|_F$.

PROOF OF THE SUFFICIENCY. Let $\|x\|_E \leq k\|Ax\|_F$ for all $x \in D(A)$.

Then given any $f \in E^*$, define a functional ϕ on $R(A)$ by $\phi(y) = f(x)$, where $y = Ax$. Recall that the correct solvability guarantees the unique solvability of the equation $Ax = y$, and hence x is uniquely defined by $y \in R(A)$, and so is $f(x)$. The additivity and homogeneity of ϕ is now plain.

Furthermore, the functional ϕ is bounded on $R(A)$ because

$$|\phi(y)| = |f(x)| \leqq \|f\|_{E^*}\|x\|_E \leqq k\|f\|_{E^*}\|Ax\|_F = k\|f\|_{E^*}\|y\|_F.$$

Thus $\phi(y)$ can be extended by continuity to $\overline{R(A)}$, and then to the entire space F, using the Hahn-Banach theorem (and the norm is preserved). Denote by g the resulting functional. As a result of its construction, $g(Ax) = \phi(Ax) = f(x)$ for all $x \in \mathcal{D}(A)$. In other words, $g \in \mathcal{D}(A^*)$ and $A^*g = f$, and so equation (A^*) is everywhere solvable.

OBSERVATION 3.5. The proof also points up one of the reasons why equation (A^*) might be not uniquely solvable; namely, it might be possible to extend the functional ϕ from $\overline{R(A)}$ to the entire space in several distinct ways.

THEOREM 3.6. *Equation* (A) *is everywhere solvable if and only if equation* (A^*) *is correctly solvable on* $R(A^*)$.

PROOF. Assume that every $y \in F$ can be expressed as $y = Ax$ ($x \in \mathcal{D}(A)$). If $g \in \mathcal{D}(A^*)$, then

$$|g(y)| = |g(Ax)| = |(A^*g)(x)| \leqq \|A^*g\|_{E^*}\|x\|_E.$$

For a fixed g, the functional $T_g(y) = |g(y)|/\|A^*g\|_{E^*}$ is convex and continuous on F. For a fixed y, these functionals are uniformly bounded:

$$T_g(y) \leqq \|x\|_E.$$

The uniform boundedness principle shows that

$$T_g(y) \leqq k\|y\|_F, \quad \text{or} \quad |g(y)| \leqq k\|A^*g\|_{E^*}\|y\|_F.$$

By the definition of the norm of the functional g, $\|g\|_{F^*} \leqq k\|A^*g\|_{E^*}$, which completes the proof of the theorem.

THEOREM 3.7. *If equation* (A) *is normally solvable, then equation* (A^*) *is closedly solvable.*

PROOF. Let $R(A)$ be closed. If $f_n \in R(A^*)$ and $f_n \longrightarrow f$, $f \in E^*$, then $f_n = A^*g_n$ ($g_n \in F^*$) and $f_n(x) = g_n(Ax) \longrightarrow f(x)$ for all $x \in \mathcal{D}(A)$. Therefore, the restrictions of the functionals g_n to the Banach space $R(A)$ form a sequence of bounded functionals on $R(A)$ which converges weakly to some functional $\phi(y)$, and ϕ is also a bounded linear functional on $R(A)$. By the Hahn-Banach theorem, $\phi(y)$ can be extended to a bounded linear functional $g(y)$ on the entire space F. By construction, $g(Ax) = \phi(Ax) = f(x)$, i.e., $g \in \mathcal{D}(A^*)$ and $f = A^*g$.

The theorem is proven.

The converse statement is not true for an arbitrary operator (see example on page 14). However, the fact that the adjoint equation is closedly solvable is equivalent to a specific property of the initial equation. More precisely, we have

PROPOSITION 3.8. *Equation* (A^*) *is closedly solvable if and only if there exists a constant* $k_1 > 0$ *such that the set* R_1 *of all* $y \in R(A)$ *having the property that equation* (A) *has a solution* x *which satisfies* $\|x\|_E \leq k_1 \|y\|_F$ *is dense in* $R(A)$.

PROOF OF NECESSITY. The operator A^* is closed; therefore, if equation (A^*) is closedly solvable, Theorem 2.3 guarantees that given any $f \in R(A^*)$, there exists $g \in \mathcal{D}(A^*)$ such that $A^*g = f$ and $\|g\| \leq k\|f\|$, and k does not depend on the choice of f.

Let us show that the statement of the proposition is true for $k_1 > k$. Assuming the converse, we can find $y_0 \in R(A)$ and a neigborhood of y_0 such that given any y in this neighborhood, all its preimages x under A satisfy $\|x\| > k_1\|y\|$. With no loss of generality, one can take $\|y_0\| = 1$, and so all preimages of the elements in a neighborhood of y_0 sit outside the ball $\|x\| \leq k$. Let $\mathcal{D}_k$ denote the intersection of this ball with $\mathcal{D}(A)$. The set $\overline{A\mathcal{D}_k}$ is closed, convex, and balanced in F, and $y_0 \notin \overline{A\mathcal{D}_k}$. Consequently, there is a linear functional $g_1 \in F^*$ such that

$$g_1(y_0) > \sup_{x \in \mathcal{D}_k} |g_1(Ax)|.$$

In particular, this implies that the functional $g_1(Ax)$ is bounded on $\mathcal{D}(A)$ (its norm is not larger than $g_1(y_0)/k$), i.e., $g_1 \in \mathcal{D}(A^*)$.

Let $f = A^*g_1$ and now find a functional $g \in \mathcal{D}(A^*)$ such that $f = A^*g$ and $\|g\| \leq k\|A^*g\|$. Since $A^*(g - g_1) = 0$, the functionals g and g_1 are identical on $R(A)$. We obtain

$$g(y_0) > \sup_{x \in \mathcal{D}_k} |g(Ax)| = \sup_{x \in \mathcal{D}_k} |(A^*g)(x)| = k\|A^*g\| \geq \|g\| .$$

This implies that $\|y_0\| > 1$, which contradicts our choice $\|y_0\| = 1$. Therefore, the condition is necessary.

PROOF OF SUFFICIENCY. Let the sequence of functionals A^*g_n converge in E^* to $f \in E^*$. Then

$$g_n(Ax) \longrightarrow f(x) \quad (x \in \mathcal{D}(A)) \quad \text{and} \quad \|A^*g_n\| \le M.$$

For any $y \in R_1$, we have

$$|g_n(y)| = |(A^*g_n)(x)| \le M\|x\| \le Mk_1\|y\|.$$

Therefore, the sequence of the restrictions of the g_n to R_1 is bounded on the set R_1 and converges here. Since R_1 is dense in the space $\overline{R(A)}$, this sequence converges on the entire space $\overline{R(A)}$ to some bounded linear functional, which can be extended in turn to a bounded linear functional g on the whole of F. On $\mathcal{D}(A)$ we have

$$g_n(Ax) \longrightarrow g(Ax) = f(x),$$

i.e., $g \in \mathcal{D}(A^*)$ and $f = A^*g$. Thus, the set $R(A^*)$ is closed, which completes the proof of the proposition.

To summarize, we shall write the main results of this section in the following table.

For an equation with a densely defined operator

(A)		(A*)
unique solv.	⇐	*dense solv.*
dense solv.	⇔	*unique solv.*
correct solv.	⇔	*everywhere solv.*
everywhere solv.	⇒	*correct solv.*
normal solv.	⇒	*closed solv.*

Usually, it is relatively easy to establish the dense solvability of an equation. This is why a certain method of proving uniqueness for differential equations is common: namely, it suffices to show that the adjoint equation is densely solvable. For proving existence theorems, the above results are not very useful, because they only allow us to draw conclusions about the everywhere or closed solvability of the equation adjoint to the given one.

Now we shall present an example which shows that in the above table, the arrows pointing only in one direction cannot be reversed in general. If E is an infinite dimensional Banach space, then using the existence of an algebraic (Hammel) basis, one can easily construct a linear functional $\phi(x)$ defined on the entire space E, but not continuous. The kernel $N(\phi)$ of this functional is dense in E. Consider the operator A from E into E given by

$$Ax = x - \phi(x)u,$$

where u is a fixed element of E. There are two possibilities:

1) if $\phi(u) \neq 1$, then equation $Ax = y$ is everywhere uniquely solvable, and $x = y + [\phi(y)/(1 - \phi(y))]u$;

2) if $\phi(u) = 1$, then equation $Ax = y$ is solvable only for $y \in N(\phi)$, and we get $x = y$.

In the second case, $N(A) = \{cu\}$.

Now consider the operator A^* adjoint to A. If $g \in E^*$, then

$$g(Ax) = g(x - \phi(x)u) = g(x) - \phi(x)g(u).$$

Since the functional $\phi(x)$ is not bounded, $g(Ax)$ is a bounded functional if and only if $g(u) = 0$. Therefore, $\mathcal{D}(A^*) = \{g : g(u) = 0\}$ is not dense in E^*. Furthermore, $A^*g = g$ for every $g \in \mathcal{D}(A^*)$, and so $R(A^*)$ is a hyperplane in E^*.

In our example, we see that, in case 1) the unique solvability of (A) does not imply the dense solvability of (A*). In case 2), the correct and closed solvability of equation (A*) does not imply normal solvability and so certainly does not imply everywhere solvability of (A). (We remark that in case 1), the kernel $N(A)$ is not an orthogonal complement of $R(A^*)$.)

§4. THE EQUATION ADJOINT TO THE FACTORED EQUATION

Assume that $N(A)$ is closed and consider the operator $\tilde{A}^*$, the adjoint to the operator $\tilde{A}$ from $\tilde{E} = E/N(A)$ to F. Then $\tilde{A}^*$ is defined on $\mathcal{D}(\tilde{A}^*) \subset F^*$ and takes its value in $\tilde{E}^*$. Using the general theorems, $\tilde{E}^* = N(A)^{\perp}$, where the orthogonal complement is taken in E^*. Let us show that $\mathcal{D}(\tilde{A}^*)$ and $\mathcal{D}(A^*)$ are the same. Indeed, let $g \in \mathcal{D}(A^*) \subset F^*$, i.e., the functional $g(Ax)$ is bounded on $\mathcal{D}(A)$:

$$|g(Ax)| \leq c\|x\|_E \qquad (x \in \mathcal{D}(A)).$$

If X is the coset of x, then, by definition, $\tilde{A}X = Ax$, and hence $|g(\tilde{A}X)| \leq c\|x\|_E$. We obtain

$$|g(\tilde{A}x)| \leq c \inf_{x \in X} \|x\|_E = c\|X\|_{\tilde{E}},$$

i.e., $g \in \mathcal{D}(\tilde{A}^*)$. Conversely, if $g \in \mathcal{D}(\tilde{A}^*)$, then

$$|g(Ax)| = |g(\tilde{A}X)| \leq c\|X\|_{\tilde{E}} \leq c\|x\|_E,$$

i.e., $g \in \mathcal{D}(A^*)$.

The functionals $\tilde{f} = \tilde{A}^*g$ and $f = A^*g$ are related through

$$\tilde{f}(X) = g(\tilde{A}X) = g(Ax) = f(x)$$

for all $x \in \mathcal{D}(A)$. Since $\mathcal{D}(A)$ and $\mathcal{D}(\tilde{A})$ are dense in E and $\tilde{E}$ respectively, $\tilde{f}(X) = f(x)$ for all $X \in \tilde{E}$ and $x \in E$. But the last relation gives us an isometric correspondence between the spaces $\tilde{E}^* = (E/N(A))^*$ and $N(A)^{\perp}$, and, modulo this isometry, we can say that $R(A^*)$ and $R(\tilde{A}^*)$ are also identical. Now using Theorem 3.4 we obtain directly

THEOREM 4.1. *Let* $N(A)$ *be closed. Then equation* $(\tilde{A})$ *is correctly solvable if and only if equation* (A^*) *is normally solvable.*

§ 5. AN EQUATION WITH A CLOSED OPERATOR WHICH HAS A DENSE DOMAIN

Let A be a closed operator. Then the following important theorem is true.

THEOREM 5.1. *If* A *is closed and equation* (A^*) *is closedly solvable, then equation* (A) *is normally solvable.*

PROOF. We shall show that $R(A)$ is closed. Choose $y \in \overline{R(A)}$. Then there is $x_o \in \mathcal{D}(A)$ such that $\| y - Ax_o \|_F < 1$. Using Proposition 3.8, one can find an element $y_1 = Ax_1 \in R_1$ satisfying

$$\| y - Ax_o - Ax_1 \|_F < 1/2 .$$

Since R_1 consists of rays originating from θ, we can assume that $\| y_1 \|_F < 1$, and then $\| x_1 \|_E < k_1$. Now pick $y_2 = Ax_2 \in R_1$ such that

$$\| y - A(x_o + x_1 + x_2) \|_F < 1/4 \quad \text{and} \quad \| x_2 \|_E < k_1/2 .$$

Continuing the process, we produce a sequence x_n, $n = 1,2,\dots$, satisfying

$$\| y - A(x_o + x_1 + \dots + x_n) \|_F < 1/2^n \quad \text{and} \quad \| x_n \|_E < k_1/2^{n-1}.$$

Since E is complete, the series $x_o + x_1 + \dots + x_n + \dots$ converges to an element x; thus

$$\lim_{n\to\infty} (x_o + x_1 + \dots + x_n) = x$$

and

$$\lim_{n\to\infty} (Ax_o + Ax_1 + \dots + Ax_n) = y.$$

Because A is closed, $x \in \mathcal{D}(A)$ and $y = Ax$, i.e., $y \in R(A)$.

The theorem is proven

Recall that for A closed, the normal solvability of equation (A) is equivalent to the correct solvability of equation ($\tilde{A}$). Applying Theorem 4.1, the latter implies the normal solvability of equation (A^*). In other words, we have

THEOREM 5.2. *If* A *is a closed operator, then equation* (A*) *is closedly solvable if and only if it is normally solvable.*

Assume now that equation (A*) is correctly solvable. By Theorem 2.1, this means that (A*) is uniquely and closedly solvable. The first property implies (Theorem 3.2) that (A) is densely solvable, while the second guarantees that (A) is normally sovable. Therefore, $E = \overline{R(A)} = R(A)$, i.e., (A) is everywhere solvable.

THEOREM 5.3. *Let* A *be closed. Then equation* (A) *is everywhere solvable if and only if equation* (A*) *is correctly solvable.*

For equations with closed operators, the table on page 14 becomes more complete.

If A *is a closed operator with dense domain, then*

(A)		(A*)		
unique solv.	$\Leftarrow$	*dense solv.*		
dense solv.	$\Leftrightarrow$	*unique solv.*		
correct solv.	$\Leftrightarrow$	*everywhere solv.*		
normal solv.	$\Leftrightarrow$	*closed solv.*	$\equiv$	*normal solv.*
everywhere solv.	$\Leftrightarrow$	*correct solv.*		

In general, the uppermost arrow cannot be reversed, even when A is bounded. To provide an example, let $E = F = L_1[0,1]$ and consider the bounded operator A in $L_1[0,1]$ given by

$$Ax(t) = \int_0^t x(s)ds$$

Obviously, $N(A) = \theta$. The dual space is $F^* = E^* = L_\infty[0,1]$. Let us find the adjoint operator A^*. We have

$$\int_0^1 Ax(t)g(t)dt = \int_0^1\int_0^t x(s)g(t)ds\,dt = \int_0^1 x(s)\int_s^1 g(t)dt\,ds = $$

$$= \int_0^1 x(s)A^*g(s)ds.$$

Therefore,

$$A^*g(s) = \int_s^1 g(t)dt .$$

The range $R(A^*)$ consists of the functions $f(s)$ which satisfy a Lipschitz condition together with the condition that $f(1) = 0$. The distance from $R(A^*)$ to the function $f_0(s) \equiv 1$ is equal to 1, and so $R(A^*)$ is not dense in E^*.

Consequently, equation (A) in our example is uniquely solvable, while equation (A*) is not densely solvable.

Finally, we make a few additional remarks.

THE GRAPH METHOD. In the direct product $E \times F$ take the subset Γ_A of all pairs $\{x,Ax\}$ with $x \in \mathcal{D}(A)$. This linear manifold is called the *graph* of the operator A. It follows from the definition that A is a closed operator if and only if its graph is a subspace of the space $E \times F$. The space dual to $E \times F$ is $E^* \times F^*$. Now assuming that $\mathcal{D}(A)$ is dense in E, consider the orthogonal complement $\Gamma_A^{\perp}$ of the graph Γ_A in $E^* \times F^*$, i.e., the set of all pairs $\{f,g\}$ ($f \in E^*$, $g \in F^*$) such that $f(x) + g(Ax) = 0$ for all $x \in \mathcal{D}(A)$. The equality $g(Ax) = - f(x)$ shows that $g \in \mathcal{D}(A^*)$ and $-A^*g = f$. Conversely, the last relation implies that $\{f,g\} \perp \Gamma_A$. We formulate this remark as

PROPOSITION 5.4. *The graph* Γ_{-A^*} *of the operator* $-A^*$ *is the orthogonal complement in* $E^* \times F^*$ *of the graph of* A.

Indeed, we have shown that the orthogonal complement $\Gamma_A^{\perp}$ consists of all the pairs $\{-A^*g,g\}$ with $g \in \mathcal{D}(A^*)$.

We remark that, in particular, Proposition 5.4. implies the fact that A^* is closed — a result which we already proved.

PROPOSITION 5.5. *If the operator* A *has a dense domain in* E *and is closed, then* $\mathcal{D}(A^*)$ *is a total set, i.e., given any* $y_0 \in F$, *there is a functional* $g_0 \in \mathcal{D}(A^*)$ *such that* $g_0(y_0) \neq 0$.

PROOF. It is clear that the element $\{\theta,y_0\}$ is not in the graph Γ_A. Therefore, in $E^* \times F^*$ one can find a functional which is orthogonal to Γ_A and is not zero at $\{\theta,y_0\}$. But we know that such a functional has the form $\{-A^*g_0,g_0\}$, where $g_0 \in \mathcal{D}(A^*)$. At the same time

$$-A^*g_0(\theta) + g_0(y_0) \neq 0, \quad \text{i.e.,} \quad g_0(y_0) \neq 0 .$$

The proposition is proven.

Recall that a total linear set is always dense in F^* in the weak topology, and so given a closed operator, the domain of its adjoint cannot be too sparse.

THE CASE OF REFLEXIVE SPACES. In the dual E^* of a reflexive space, every linear weakly dense set is dense. We obtain

PROPOSITION 5.6. *If* F *is reflexive and* A *is closed and has a dense domain, then* A^* *has also a dense domain in* F^*.

We conclude by showing that for reflexive spaces the first implication in the table for this section may be reversed.

THEOREM 5.7. *If the space* E *is reflexive and the operator* A *is closed and has a dense domain, then the unique solvability of equation* (A) *implies the dense solvability of equation* (A^*).

PROOF. We consider the operator A^{-1}, which can be defined, due to the unique solvability of (A), as an operator from the Banach space $\overline{R(A)}$ to the Banach space E. Since the domain of A^{-1} is dense in $\overline{R(A)}$ and E is reflexive, Proposition 5.6. shows that the domain $\mathcal{D}((A^{-1})^*)$ is dense in E^*. One may verify the equality $\mathcal{D}((A^{-1})^*) = \mathcal{D}(A^*)$ directly and this completes the proof of the theorem.

§ 6. NORMALLY SOLVABLE EQUATIONS WITH FINITE DIMENSIONAL KERNEL

A subset M of the Banach space E is called *locally compact* in E if the intersection of M with any ball of E is compact in E.

THEOREM 6.1. *Let* A *be a closed operator. In order that the equation* (A) *be normally sovable and that the homogeneous equation* $Ax = \theta$ *have only a finite number of linearly independent solutions, it is necessary and sufficient that the preimage of each compact subset of* $R(A)$ *under* A *should be locally compact.*

PROOF OF THE SUFFICIENCY. The kernel $N(A)$ of A, being the preimage of the element $y = \theta$, is locally compact, and hence finite dimensional (any linear, normed, locally compact space is finite dimensional). Furthermore, the preimage of any element $Ax = y \in R(A)$ consists of the elements $x + z$, where $z \in N(A)$ is arbitrary. Since $N(A)$ is finite dimensional, there is an element $\tilde{x}$ of the form $x + z$ with the smallest norm: $\|\tilde{x}\| = \inf_{z \in N(A)} \|x + z\|$. We show that there is a constant $k > 0$ not depending upon the choice of $y \in R(A)$, and such that

$$\|\tilde{x}\| \leq k\|y\|, \quad y = A\tilde{x} .$$

Suppose that there is no such k. Then one can find a sequence $y_n = A\tilde{x}_n$ satisfying

$$\frac{\|\tilde{x}_n\|}{\|A\tilde{x}_n\|} \longrightarrow \infty .$$

Set $u_n = \tilde{x}_n/\|\tilde{x}_n\|$ and $v_n = Au_n$. Then $\|u_n\| = 1$ and $\|Au_n\| = \|v_n\| \longrightarrow 0$. According to the hypothesis, there are both a subsequence n' and an element u such that $u_{n'} \longrightarrow u$, and since A is closed, $Au = 0$, i.e., $u \in N(A)$. Obviously, the element $u_{n'}$ has the smallest norm in the preimage of $v_{n'}$, and the equality $A(u_{n'} - u) = Au_{n'} = v_{n'}$ implies

$$\|u_{n'} - u\| \geq \|u_{n'}\| = 1 .$$

This contradicts our assumption that $u_{n'} \longrightarrow u$.

Therefore, the constant k exists, and one may apply Theorem 2.3. to obtain the normal solvability of (A).

PROOF OF THE NECESSITY. Now assume that (A) is normally solvable and that $N(A)$ is finite dimensional. Let Q be a compact subset of $R(A)$ and let M be its preimage under A. Now choose a sequence $x_n \in M$ and contained in the ball $\|x\| \leq c$. Since Q is compact, there exists a subsequence n' such that $Ax_{n'} \longrightarrow y$. Furthermore, one can find a sequence $\bar{x}_{n'} \in D(A)$ such that $\bar{x}_{n'} \longrightarrow x$ and $A\bar{x}_{n'} = Ax_{n'}$, and we may write $x_{n'} = \bar{x}_{n'} + z_{n'}$, with $z_{n'} \in N(A)$. The sequence $z_{n'}$ is bounded, because $x_{n'}$ are bounded and $\bar{x}_{n'}$ converges. Since $N(A)$ is finite dimensional, one can take a convergent subsequence $z_{n''}$. Then the subsequence $x_{n''} = \bar{x}_{n''} + z_{n''}$ converges, which proves the local compactness of M and completes the proof of the theorem.

From now on, a normally solvable equation given by a closed operator A with a finite dimensional kernel will be called n-*normal*, and we shall denote the dimension of $N(A)$ by $n(A)$ (alternatively, $n(A) = \dim \operatorname{Ker} A$).

The n-normal equations have the following property:

If equation (A) *is* n-*normal and* B *is an arbitrary compact operator from* E *to* F, *then the equation* $Ax + Bx = y$ *is also* n-*normal*.

Indeed, let Q be a compact subset of F, and let M_1 denote the preimage of Q under $A + B$. If $x_n \in M_1$ and $\|x_n\| \leq c$, then since B is compact, one can find a subsequence n' such that $Bx_{n'}$ converges. The sequence $y_{n'} = Ax_{n'} + Bx_{n'}$ is contained in the compact set Q and hence it has a convergent subsequence $y_{n''}$. Now the sequence $Ax_{n''} = y_{n''} - Bx_{n''}$ is also convergent and, using the n-normality of equation (A), one can extract a convergent subsequence from the sequence $x_{n''}$.

§7. A PRIORI ESTIMATES

As we have shown in § 3, if A is an arbitrary operator with a dense domain and if equation (A) is correctly solvable on $R(A)$, i.e., if one has the estimate

$$\| x \|_E \leq k \| Ax \|_F \quad (x \in \mathcal{D}(A)), \tag{7.1}$$

then the adjoint equation (A^*) is everywhere solvable. To obtaint the estimate (7.1) one need not know for which right-hand sides equation (A) is solvable. Looking from the point of view of the theory of equations, the content of (7.1) is : if x is any solution of equation (A), the it satisfies $\| x \|_E \leq k \| y \|_F$. This is the reason why such estimates are known under the name of *a priori estimates*.

In §5 we proved that for a closed operator A, the existence of the a priori estimate

$$\| g \|_{F^*} \leq k \| A^* g \|_{E^*} \quad (g \in \mathcal{D}(A^*)) \tag{7.2}$$

implies that equation (A) is everywhere solvable.

Therefore, one can prove existence theorems for solutions of linear equations using a priori estimates. *If* A *is closed, then the a priori estimates* (7.1) *and* (7.2) *are necessary and sufficient conditions for the everywhere solvability of the equations* (A) *and* (A^*).

However, we should remark that it is sometimes difficult to obtain estimate (7.2) since one has to prove it for all $g \in \mathcal{D}(A^*)$, or at least for all g in a subset M of $\mathcal{D}(A^*)$ such that A^* is the closure of the restriction of A^* to M. Describing the entire domain of definition of the adjoint operator is not always an easy task.

In the theory of elliptic differential equations, one is often able to get a priori estimates which are weaker than (7.1), in the following setting: let the space E be embedded in some larger space E_0, i.e., if $x \in E$, then $x \in E_0$ and $\| x \|_{E_0} \leq c \| x \|_E$.

The space E is said to be *compactly embedded* in the space E_0 if every bounded subset of E is a compact subset of E_0.

Let A be a closed operator from E to F, with domain $\mathcal{D}(A) \subset E$. The a priori estimate that we now are discussing has the form

$$\|x\|_E \leq a\|x\|_{E_0} + b\|Ax\|_F \quad (x \in \mathcal{D}(A)). \tag{7.3}$$

THEOREM 7.1. *If* E *is compactly embedded in* E_0 *and* A *is a closed operator, then equation* (A) *is* n-*normal if and only if estimate* (7.3) *holds.*

PROOF OF THE NECESSITY. If (A) is normally solvable, then given any $x \in \mathcal{D}(A)$ there is $\tilde{x} \in \mathcal{D}(A)$ such that $A\tilde{x} = Ax$ and

$$\|\tilde{x}\|_E \leq k_1\|Ax\|_F .$$

Now $x = \tilde{x} + z$, where $z \in N(A)$, and so

$$\|x\|_E \leq k_1\|Ax\|_F + \|z\|_E .$$

In the finite dimensional space $N(A)$ all norms are equivalent and so

$$\|x\|_E \leq k_1\|Ax\|_F + k_2\|z\|_{E_0} \leq k_1\|Ax\|_F + k_2\|x\|_{E_0} + k_2\|\tilde{x}\|_{E_0} \leq$$

$$\leq k_1\|Ax\|_F + k_2\|x\|_{E_0} + ck_2\|\tilde{x}\|_E \leq (k_1 + ck_1k_2)\|Ax\|_F +$$

$$+ k_2\|x\|_{E_0} \leq a\|x\|_{E_0} + b\|Ax\|_F .$$

PROOF OF THE SUFFICIENCY. Consider A as an operator from E_0 to F with domain $\mathcal{D}(A)$. This new operator is also closed: if $x_n \longrightarrow x$ in E_0 and $Ax_n \longrightarrow y$ in F, then (7.3) implies

$$\|x_n - x_m\|_E \leq a\|x_n - x_m\|_{E_0} + b\|Ax_n - Ax_m\|_F \longrightarrow 0$$

as $n,m \longrightarrow \infty$. Since E is complete, there is $\bar{x} \in E \subset E_0$ such that $x_n \longrightarrow \bar{x}$ in E. But $\|\bar{x} - x_n\|_{E_0} \leq c\|\bar{x} - x_n\|_E \longrightarrow 0$, i.e., $\bar{x} = x$. Because A is closed as an operator from E to F, we get $Ax = y$.

Now let $\{Ax\}$ be a compact subset of F, and assume that $\forall\, x$, $\|x\|_{E_0} \leq c$. Then $\{Ax\}$ is bounded in F and so, by (7.3), the set $\{x\}$ is bounded in E. But E is compactly embedded in E_0, which implies that $\{x\}$ is compact in E_0. According to Theorem 6.1, equation (A) is n-normal.

The theorem is proven.

Let A be a closed operator with domain $\mathcal{D}(A) \subset E_0$ and values in F. As was shown in §2, $\mathcal{D}(A)$ is a Banach space E_{0A} relative to the norm

$$\|x\|_{E_{0A}} = \|x\|_{E_0} + \|Ax\|_F .$$

If one sets $E = E_{0A}$ in the above arguments, then inequality (7.3) is obviously satisfied (in fact, it becomes an equality with $a = b = 1$).

We obtain

THEOREM 7.2. *Let the operator* A *with domain* $\mathcal{D}(A) \subset E_0$ *and values in* F *be closed. For the equation* (A) *to be* n-*normal it is sufficient that the space* E_{0A} *be compactly embedded in the space* E.

Obviously, the converse is not true.

§8. EQUATIONS WITH FINITE DEFECT

Let F_1 be a subspace of the Banach space F. The dimension of the orthogonal complement $F_1^{\perp}$ is called the *defect* or the *codimension* of the subspace F_1:

$$\operatorname{def} F_1 = \dim F_1^{\perp} .$$

This number can be finite or infinite. On the assumption that $\operatorname{def} F_1 = n < \infty$, let $g_1,\ldots,g_n$ be a basis for $F_1^{\perp}$. If $u_1 \notin F_1$, then one can construct a linear functional $\phi_1 \in F_1^*$ such that $\phi_1(u_1) = 1$ and $\phi_1(y) = 0$ for all $y \in F_1$. Then $\phi_1 \in F_1^{\perp}$. Now let u_2 be an element which is not in the linear span of the subspace F_1 together with the element u_1 (which is again a subspace of F), and choose $\phi_2 \in F_1^{\perp}$ such that $\phi_2(u_2) = 1$, $\phi_2(u_1) = 0$. This process may be continued, and clearly it will have at most n steps: the functionals ϕ_i are linearly independent (if $\sum c_i\phi_i(z) = 0$ for all $z \in F$, then taking successively $z = u_1$, $z = u_2,\ldots$, we obtain $c_1 = 0$, $c_2 = 0,\ldots$) and are all in $F_1^{\perp}$. In fact, the number of steps is exactly n. Indeed, if the process terminates at the m-th step, then one can write every element $z \in F$ as

$$z = \sum_{k=1}^{m} \alpha_k u_k + y \qquad (y \in F_1).$$

For $m < n$, the system of linear equations

$$\sum_{i=1}^{n} c_i g_i(u_j) = 0 \qquad (j = 1,2,\ldots,m)$$

has a nontrivial solution. Taking the above representation of the elements of F, this implies that the functional $\sum_{i=1}^{n} c_i g_i$ is zero on all $z \in F$, which is impossible because $\{g_i\}$ is a basis of $F_1^{\perp}$.

Therefore, $m = n$ and F decomposes into the direct sum of its subspace F_1 and the n-dimensional space L_n spanned by the elements $u_1,\ldots,u_n$.

We have recalled this familiar argument with a specific purpose in mind. If $\mathcal{D}$ is a linear set dense in F, then one can choose the $u_1,\ldots,u_n$ above to be elements of $\mathcal{D}$. Namely, first take $\bar{u}_1 \notin F_1$

arbitrary, and then approximate $\bar{u}_1$ by an element $u_1 \in \mathcal{D}$ such that $\| \bar{u}_1 - u_1 \|_F$ is less than the distance from u_1 to F_1. Then $u_1 \in \mathcal{D}$ and $u_1 \notin F_1$. Now proceed in the same way at each step.

If $u_k \in \mathcal{D}$ $(k = 1,2,\ldots,n)$, then the intersection $\mathcal{D} \cap F_1$ is dense in F_1. Indeed, let $y \in F_1$. Then there exists a sequence $z_s \in \mathcal{D}$ such that $z_s \longrightarrow y$. Write $z_s = v_s + y_s$, with $v_s \in L_n$ and $y_s \in F_1$. Since $v_s \in \mathcal{D}$, we have $y_s \in \mathcal{D}$, and so $y_s \in \mathcal{D} \cap F_1$. Now $y - z_s = y - y_s - v_s \longrightarrow 0$ implies $y - y_s \longrightarrow 0$, i.e., $\mathcal{D} \cap F_1$ is dense in F_1.

Let us summarize our result.

LEMMA 8.1. *Let* F *be a subspace of* F *having a finite defect* n *and let* $\mathcal{D}$ *be a dense linear set in* F. *Then* F *decomposes into the direct sum* $F = F_1 \oplus L_n$, *where* $\dim L_n = n$, $L_n \subset \mathcal{D}$, *and the intersection* $\mathcal{D} \cap F_1$ *is dense in* F_1.

Let A be an operator from a Banach space E to a Banach space F. The *defect* $d(A)$ *of equation* (A) (or *of the operator* A) is the defect of the subspace $\overline{R(A)}$ in F.

If $\mathcal{D}(A)$ is dense in E, then the adjoint A^* exists and applying Proposition 3.1,

$$\operatorname{def} \overline{R(A)} = \dim R(A)^{\perp} = \dim N(A^*). \tag{8.1}$$

If A is closed, we say that equation (A) is d-*normal* if it is normally sovable and has a finite defect $(d(A) < \infty)$. [In this case, the explicit notation $\dim \operatorname{Coker} A$ is frequently used for $d(A)$.]

Recalling Theorem 2.4, in order to check that an equation with a closed operator is d-normal, it suffices to show that $R(A)$ has a finite dimensional direct complement in F.

If (A) is d-normal and $\overline{\mathcal{D}(A)} = E$, then the adjoint equation (A*) is normally solvable (Theorems 3.7 and 5.2) and has a finite dimensional kernel; in other words, equation (A*) is n-normal. By Theorem 5.1 and Lemma 8.1, the converse is also true when A is closed. Therefore

If the operator A *is closed and* $\overline{\mathcal{D}(A)} = E$, *then equation* (A) *is* d-*normal if and only if equation* (A*) *is* n-*normal.*

Now if equation (A) is n-normal, then equation (A*) is normally solvable, i.e., $R(A^*) = N(A)^{\perp}$, and since $N(A)$ is finite dimensional, $\operatorname{def} R(A^*) = \dim N(A) < \infty$.

Conversely, if A is closed and equation (A*) is d-normal, then, according to Theorems 3.7 and 5.1, equation (A) is normally solvable and

$$\dim N(A) \leq \operatorname{def} N(A)^{\perp} = \operatorname{def} R(A^*) < \infty .$$

Thus $N(A)$ is finite dimensional and then

$$\dim N(A) = \operatorname{def} N(A)^{\perp} = \operatorname{def} R(A^*).$$

THEOREM 8.2. *Let* A *be a closed operator with* $\overline{D(A)} = E$. *Then equation* (A) *is* n-*normal* (*respectively,* d-*normal*) *if and only if equation* (A*) *is* d-*normal* (*respectively,* n-*normal*).

Applying Theorem 7.1 to equation (A*), we obtain the following criterion for the d-normality of equation (A).

THEOREM 8.3. *Let* A *be a closed operator with* $\overline{D(A)} = E$ *and suppose that* F^* *is compactly embedded in a Banach space* G. *Then equation* (A) *is* d-*normal if and only if the a priori estimate*

$$\| g \|_{F^*} \leq a \| g \|_G + b \| A^* g \|_{E^*} \qquad (g \in D(A^*))$$

holds.

§ 9. SOME DIFFERENT ADJOINT EQUATIONS

Let us assume that the space E is embedded in a space E_0 as a dense subset. Then E_0^* is embedded in the space E^*. Indeed, every linear functional $g \in E_0^*$ is naturally defined on E and is bounded:

$$|g(x)| \leq \|g\|_{E_0^*} \|x\|_{E_0} \leq c\|g\|_{E_0^*} \|x\|_E \qquad (x \in E).$$

Thus $g \in E^*$ and $\|g\|_{E^*} \leq c\|g\|_{E_0^*}$.

Moreover, if $\|g\|_{E^*} = 0$, then $g(x) \equiv 0$ on E and hence $g(x) \equiv 0$ on E_0. We conclude that the constructed map $E_0^* \longrightarrow E^*$ is one-to-one.

Let A be alinear operator from E to F such that $\mathcal{D}(A)$ is dense in E. Frequently, in applications, the space F^* is not easy to describe, and this is the reason why another adjoint operator is considered instead of the adjoint operator A^* acting from F^* to E^*. Namely, one can look at A as an operator from E_0 to F, with a dense domain $\mathcal{D}(A) \subset E_0$. The adjoint $\hat{A}^*$ of this operator already acts from F^* to E_0^*. To clarify the connection between A^* and $\hat{A}^*$, let $g \in F^*$ be an element of $\mathcal{D}(\hat{A}^*)$. Then

$$|g(Ax)| \leq c_1\|x\|_{E_0} \leq c_2\|x\|_E \qquad (x \in \mathcal{D}(A) \subset E),$$

and hence $g \in \mathcal{D}(A^*)$. Furthermore, the functionals $\hat{A}^*g(x)$ and $A^*g(x)$ coincide on $\mathcal{D}(A)$ (they are both equal to $g(Ax)$), and since $\mathcal{D}(A)$ is dense in E_0, A^*g extends by continuity to $\hat{A}^*g$. In other words, the operator A^* is an extension of the operator $\hat{A}^*$. We remark that if $A^*g \in E_0^*$, then

$$|g(Ax)| = |A^*g(x)| \leq c_1\|x\|_{E_0},$$

and so $g \in \mathcal{D}(\hat{A}^*)$, i.e., $\hat{A}^*$ is the restriction of A^* to those functionals whose images are in E_0^*.

In particular, we see that *the operators* A^* *and* $\hat{A}^*$ *have the same kernel.*

Now assume that $R(A^*)$ is closed in E^* and let us show that the same holds true for $R(\hat{A}^*)$. Take $f_n \in R(\hat{A}^*)$ and let $f_n \longrightarrow f$ in E_0^*.

Then $f_n \in R(A^*) \cap E_0^*$, and the inequality $\|g\|_{E^*} \leq c\|g\|_{E_0^*}$ $(g \in E_0^*)$ shows that the sequence f_n converges in E^* also. Since $R(A^*)$ is closed in E^*,

$$f \in R(A^*) \cap E_0^* = R(\hat{A}^*).$$

Therefore, *if equation* (A^*) *is closedly solvable, then so is equation* $(\hat{A}^*)$.

Moreover, *if equation* (A^*) *is normally solvable, then so is equation* $(\hat{A}^*)$. Indeed, $f \in E_0^*$ and $f \perp N(A)$ imply that $f \in E^*$ and that the equation $A^*g = f$ is solvable. But because $f \in E_0^*$, one has that $\hat{A}^*g = A^*g = f$, i.e., $f \in R(\hat{A}^*)$.

Now consider the relationship between the defects of equations (A^*) and $(\hat{A}^*)$.

If equation (A^*) *is* d-*normal, then so is equation* $(\hat{A}^*)$.

To prove this statement, decompose the space E^* into a direct sum $E^* = R(A^*) \oplus L^*$, where L^* is assumed to be finite dimensional. Denote by P the projection onto L^* defined by this decomposition. Then PE_0^* is a subspace of L^*. Let $h_1,\ldots,h_s$ be a basis of PE_0^*, and let $f_1,\ldots,f_s$ be any elements of E_0^* such that $Pf_k = h_k$, $k = 1,\ldots,s$. Given an arbitrary element $f \in E_0^*$, we can write it as $f = g + h$, with $g \in R(A^*)$ and $h = Pf = \sum_{k=1}^{s} c_k h_k$. Set

$$\hat{f} = f - \sum_{k=1}^{s} c_k f_k \in E_0^* .$$

Then

$$P\hat{f} = Pf - \sum_{k=1}^{s} c_k Pf_k = h - \sum_{k=1}^{s} c_k h_k = 0,$$

i.e., $\hat{f} \in R(A^*)$. Therefore, $\hat{f} \in R(A^*) \cap E_0^* = R(\hat{A}^*)$, and $f = \hat{f} + \sum_{k=1}^{s} c_k h_k$. In other words, E_0^* decomposes into the direct sum of $R(\hat{A}^*)$ and the linear span of the elements $f_1,\ldots,f_s$. This shows that the equation $(\hat{A}^*)$ has a finite defect and that $d(\hat{A}^*) \leq d(A^*)$.

Summing up, we obtain the following result.

THEOREM 9.1. *If the space* E *is densely embedded in a space* E_0, *then the closed solvability, normal solvability,* n-*normality, or* d-*normality of equation* $(\hat{A}^*)$ *results from the closed solvability, normal solvability,* n-*normality, or* d-*normality respectively of equation* (A^*). *Moreover, the kernels of* (A^*) *and* $(\hat{A}^*)$ *coincide, and if* $d(A^*) < \infty$, *then* $d(\hat{A}^*) \leq d(A^*)$.

Now we suppose that A is closed as an operator from E_0 to F. This implies that A is closed also as an operator from E to F:

if $x_n \longrightarrow x$ in E and $Ax_n \longrightarrow y$ in F, then $x_n \longrightarrow x$ in E_0 and since A is closed, we have $x \in \mathcal{D}(A)$ and $Ax = y$. If equation $(\hat{A}^*)$ is normally solvable, or n-normal, or d-normal, then equation (A) is normally solvable, or d-normal, or n-normal respectively. But these properties do not depend upon whether (A) is considered being in the space E or E_0. What really counts is that in both cases the operator takes values in F. Consequently, the above properties of equation (A) imply the corresponding properties of equation (A*).

THEOREM 9.2. *Let* E *be densely embedded in* E_0 *and let* A *be closed as an operator from* E_0 *to* F. *Then equation* (A*) *is normally solvable (or* n-*normal, or* d-*normal) if and only if equation* $(\hat{A}^*)$ *has the same property. In the case of* d-*normality,* $d(A^*) = d(\hat{A}^*) = n(A)$.

This theorem shows in particular that if A is closed as an operator from E_0 to F, we need consider only one of the equations (A*) or $(\hat{A}^*)$ — that one which is easier to handle — in order to study the properties of equation (A).

§ 10. LINEAR TRANSFORMATIONS OF EQUATIONS

By a linear transformation of equation (A) we mean the process of passing from (A) to a new equation

$$BAx = By \ , \tag{10.1}$$

by means of a linear operator B which acts from F into a new Banach space G. It could be that the equation

$$BAx = z \tag{BA}$$

is easier to solve than the original one. However, in dealing with such transformations we must proceed with a certain amount of caution. If the operator B is not defined everywhere on F, then the solutions of equation (A) corresponding to right-hand sides $y \notin \mathcal{D}(B)$ are not in the set of all solutions of equation (BA).

Therefore, in order to be certain that no solutions are lost, we must assume that $\mathcal{R}(A) \subset \mathcal{D}(B)$.

Further, equation (10.1) may have superfluous solutions. More precisely, if $\mathcal{N}(B) \neq \{\theta\}$, then taking $x_0 \in \mathcal{D}(A)$ and $u_0 \in \mathcal{N}(B)$, $u_0 \neq \theta$, and setting $y_0 = Ax_0 + u_0$, we see that x_0 is a solution of $BAx = By_0$, but not a solution of $Ax = y_0$.

The conclusion we have reached is: *equations* (A) *and* (10.1) *are equivalent* (*i.e., any solution of* (A) *is a solution of* (10.1) *and conversely*) *if and only if*

$$\mathcal{R}(A) \subset \mathcal{D}(B) \quad \text{and} \quad \mathcal{N}(B) = \{\theta\}.$$

We remark that the first condition is automatically fulfilled when B is a bounded operator on F.

We mentioned earlier that transforming equation (A) might be useful because equation (10.1) may be in some sense simpler than (A). However, to take advantage of this, we must know how to recover the properties of equation (A), given the properties of equation (10.1) or equation (BA). The clarification of this relationship is our present task.

Obviously, any solution of the homogeneous equation (A) is also a solution of the homogeneous equation (BA). Consequently, no linear transformation can cancel the ununiqueness of the solutions of equation (A). If the transformation B is an equivalence (as above, i.e., $\mathcal{R}(A) = \mathcal{D}(B)$ and $\mathcal{N}(B) = \{\theta\}$), then $\mathcal{N}(A) = \mathcal{N}(BA)$.

Therefore, *if equation* (BA) *is uniquely solvable, then so is equation* (A); *if* $\mathcal{N}(BA)$ *is finite dimensional, then so is* $\mathcal{N}(A)$.

When we transform equations the following important situation must be kept in mind: *even if* B *is an equivalence, an equation with a closed operator* A *might be transformed into an equation whose operator* BA *is not closed.* As an example, let A have a bounded inverse A^{-1} defined on the entire space F, take $B = A^{-1}$, and transform the equation (A). The new equation is given by the operator $BA = A^{-1}A$, which equals the identity operator on $\mathcal{D}(A)$. If $\mathcal{D}(A)$ is not closed, this operator is not closed; if $\overline{\mathcal{D}(A)} = E$, then its closure is the identity operator of E.

Frequently, the transformed operator BA is closable. In this case, one can study three transformed equations: (10.1), (BA), and

$$\overline{BA}x = z . \qquad (\overline{BA})$$

If the transformation B *is a bounded equivalence and the operator* BA *is closed, then the operator* A *is also closed.*

Indeed, let $x_n \longrightarrow x$ and $Ax_n \longrightarrow y$. Since B is bounded, $BAx_n \longrightarrow By$, and now the fact that BA is closed shows that $x \in \mathcal{D}(A)$ and $BAx = By$. But B is an equivalence and so $y = Ax$.

If the transformation B *is a bounded equivalence, and equation* (BA) *is normally solvable, then equation* (A) *is also normally solvable.*

Indeed, as above, take $Ax_n \longrightarrow y$. Then $BAx_n \longrightarrow By$ and hence there is $x \in \mathcal{D}(BA)$ such that $BAx = By$. This implies $y = Ax$.

In general, if equation $(\overline{BA})$ is normally solvable, equation (A) is not necessarily so. In the case of n-normality, the connection is stronger.

If the transformation B *is bounded,* A *is closed, and equation* $(\overline{BA})$ *is* n-*normal, then equation* (A) *is also* n-*normal.*

To see this, let $\mathcal{Q}$ be a compact subset of $\mathcal{R}(A)$ and let M be its preimage under A. Then the set $B\mathcal{Q} = S$ is also compact, and its preimage M_1 under the map $\overline{BA}$ contains its preimage under the map BA, and the latter contains M. Since M_1 is locally compact, so is M.

If the operator B *is* n-*normal and* A *is closed, then* BA *is also closed.*

Indeed, let $x_n \longrightarrow x$ and $BAx_n \longrightarrow u$. By the normal solvability of equation (B), there is a convergent sequence of elements $y_{n'} \in \mathcal{D}(B)$ such that $By_{n'} = BAx_{n'}$. Then $Ax_{n'} = y_{n'} + w_{n'}$, where $w_{n'} \in N(B)$. Now we show that the sequence $w_{n'}$ is bounded. Assume that this is not so and use the fact that $N(A)$ is finite dimensional in order to choose a subsequence n'' such that $w_{n''}/\|w_{n''}\|$ converges to w, while $\|w_{n''}\| \longrightarrow \infty$. We have

$$\frac{x_{n''}}{\|w_{n''}\|} \longrightarrow 0 \quad \text{and} \quad A\left(\frac{x_{n''}}{\|w_{n''}\|}\right) = \frac{y_{n''}}{\|w_{n''}\|} + \frac{w_{n''}}{\|w_{n''}\|} \longrightarrow w ,$$

with $\|w\| = 1$, which contradicts the fact that A is closed. Since the sequence $w_{n'}$ is bounded, one can extract a convergent subsequence $w_{n'_k}$. Then the sequence $Ax_{n'_k} = y_{n'_k} + w_{n'_k}$ converges to an element y. Finally, $y = Ax$ because A is closed, and $u = BAx$ because B is closed.

If both equations (A) *and* (B) *are* n-*normal, then so is equation* (BA).

Let S be a compact subset of G, and let M be its preimage under BA. Let $x_n \in M$, $\|x_n\| \leq c$, and set $u_n = BAx_n$. Since S is compact, there exists a convergent subsequence $u_{n'} = BAx_{n'}$. We have shown above that there is a subsequence n'_k such that $Ax_{n'_k}$ converges. Equation (A) being n-normal, the sequence $x_{n'_k}$ is compact, i.e., it has a convergent subsequence.

Now we shall look into the possibility of transforming a given equation into an equation with a bounded operator.

Assume that $N(A)$ is closed and that the subspaces $N(A)$ and $\overline{R(A)}$ have direct complements in the spaces E and F, respectively:

$$E = N(A) \oplus E_1 , \quad F = \overline{R(A)} \oplus L .$$

Let $\mathcal{D}_1 = \mathcal{D}(A) \cap E_1$ and let A_1 be the restriction of A to $\mathcal{D}_1$. Then A_1 is a one-to-one map from $\mathcal{D}_1$ onto $R(A)$ and so it has an inverse A_1^{-1} on $R(A)$. This last operator can be completed by an arbitrary linear operator defined on L, and then extended by linearity to the set $R(A) \oplus L$ which is dense in F. Denote the resulting operator by $\tilde{A}_1^{-1} = B$. If $x \in \mathcal{D}(A)$, then $x = z + v$, where $z \in N(A)$, $v \in \mathcal{D}_1$. We have

$$BAx = \tilde{A}_1^{-1}A(z + v) = \tilde{A}_1^{-1}Av = v = x - Px \quad (x \in \mathcal{D}(A)), \tag{10.2}$$

where P is the projection onto $N(A)$ defined by the decomposition

$E = N(A) \oplus E_1$. Therefore, we have transformed the given equation into an equation with a bounded operator.

If A is closed, then A_1 is also closed, and hence so is A_1^{-1}. The last operator is bounded on $R(A)$ if and only if $R(A)$ is closed, i.e., if and only if equation (A) is normally solvable. In this case the operator $\tilde{A}_1^{-1}$ can also be made bounded. To conclude: *if the operator* A *is closed, equation* (A) *is normally solvable, and the subspaces* $N(A)$ *and* $R(A)$ *have direct complements, then there is a bounded operator* B *which transforms equation* (A) *into an equation of the form* (10.2).

Conversely, *if* A *is closed and* (A) *is normally solvable, then every transformation which takes* (A) *into an equation with a bounded operator is given by an operator* B *which is bounded on* $R(A)$.

Indeed, let $BAx = Qx$, $(x \in \mathcal{D}(A))$, where Q is a bounded operator. Since A is closed and (A) is normally solvable, given any $x \in \mathcal{D}(A)$, there is x_1 such that $\|x_1\| \leq k\|Ax\|$ and $Ax = Ax_1$. Then

$$\|BAx\| = \|BAx_1\| = \|Qx_1\| \leq \|Q\|\,\|x_1\| \leq k\|Q\|\,\|Ax\| ,$$

and this shows that B is bounded on $R(A)$.

The following problem emerges: when a given normally solvable equation (A) is equivalent to an equation with a bounded operator ? In the previous construction, the operator $B = \tilde{A}_1^{-1}$ was an extension of the operator A_1^{-1} and B was defined on $R(A)$ with a trivial kernel. If $\tilde{A}_1^{-1}y = \theta$, then, writing $y = u + w$ $(u \in R(A),\ w \in L)$, we obtain

$$\tilde{A}_1^{-1}w = -\tilde{A}_1^{-1}u = -A_1^{-1}u \in \mathcal{D}_1 .$$

Therefore, the operator $\tilde{A}_1^{-1}$ takes the element $w \in L$, $w \neq \theta$, into an element of $\mathcal{D}_1$. Conversely, for $\tilde{A}_1^{-1}w \in \mathcal{D}_1$ $(w \in L)$, set $u = -A_1\tilde{A}_1^{-1}w$ to get

$$A_1^{-1}u + \tilde{A}_1^{-1}w = \tilde{A}_1^{-1}(u + w) = \theta.$$

In other words, the operator $\tilde{A}_1^{-1}$ transforms equation (A) into an equivalent equation if and only if it takes L one-to-one onto a subspace which intersects $\mathcal{D}_1$ only at zero. In particular, *if there is a bounded linear operator* C *on* L *which takes* L *one-to-one into* $N(A)$, *then there exists a bounded linear operator* B *which transforms* (A) *into an equivalent equation with a bounded operator in space* E.

transforms (A) into an equivalent equation of the form (10.2).

Indeed, take $B = A_1^{-1}$ on $R(A)$ and $B = C$ on L, and extend B linearly to the entire space F to obtain a bounded operator which

§ 11. TRANSFORMATIONS OF d-NORMAL EQUATIONS

Let equation (A) be d-normal, and let the operator B have a dense domain in F. By Lemma 8.1, one can decompose F as

$$F = R(A) \oplus L , \qquad (11.1)$$

where $\dim L = d(A)$ and $L \subset \mathcal{D}(B)$. The set of values of the operator B on $\mathcal{D}(B)$ is the range $R(B)$, while the set of its values on $R(A) \cap \mathcal{D}(B)$ is the range $R(BA)$ of the operator BA. Since

$$\mathcal{D}(B) = R(A) \cap \mathcal{D}(B) \oplus L , \qquad (11.2)$$

we have

$$R(B) = R(BA) + BL \quad .$$

However, the last sum is not necessarily direct: $R(BA)$ and BL can intersect. The subspace BL is finite dimensional, and so one can find a direct complement M to $R(BA) \cap BL$ in LB. Then

$$R(B) = R(BA) \oplus M . \qquad (11.3)$$

This decomposition shows that $R(B)$ is closed if $R(BA)$ is, i.e., *for equation* (BA) *to be normally solvable it is necessary that the transformation operator* B *be normally solvable.*

Now we compute the dimension of the subspace M. If $w \in BL \cap R(BA)$, then

$$w = BAx = Bu \quad (x \in \mathcal{D}(BA),\ \ u \in L),$$

and hence $B(Ax - u) = \theta$, i.e., $z = Ax - u \in N(B)$. Conversely, (11.2) shows that given any $z \in N(B)$, there is a unique representation $z = Ax - u$ with $u \in L \subset \mathcal{D}(B)$ and $Ax \in \mathcal{D}(B)$, i.e., $x \in \mathcal{D}(BA)$. Therefore, that part L_1 of the space L which B maps into $R(BA)$ consists of the projections on L of all elements of $N(B)$. Those elements z that go to zero under projection are precisely of the form Ax, i.e., the elements of $R(A) \cap N(B)$. We conclude that $\dim L_1$ equals the dimension of the quotient space

$$N(B)/[R(A) \cap N(B)] \quad .$$

If $N(B)$ is finite dimensional, then $\dim L_1 = n(B) - \dim R(A) \cap N(B)$. Letting L_2 denote a direct complement of L_1 in L, one can set $M = BL_2$. At the same time, (11.3) holds and B defines an isomorphism of L_2 onto M. Therefore,

$$\dim M = \dim L_2 = d(A) - \dim L_1 = d(A) - n(B) + \dim R(A) \cap N(B) \ .$$

When $R(BA)$ is closed, so is $R(B)$ and

$$\operatorname{def} R(BA) = \operatorname{def} R(B) + \dim M \quad .$$

If $R(B)$ is closed and $R(BA)$ is not, then

$$\operatorname{def} R(BA) < \operatorname{def} R(B) + \dim M \quad .$$

This implies the following statement.

THEOREM 11.1. *Let equation* (A) *be* d-*normal, and let* B *be an* n-*normal operator with a dense domain. Then* $R(BA)$ *is closed and*

$$\operatorname{def} R(BA) = \operatorname{def} R(B) + d(A) - n(B) + \dim R(A) \cap N(B) \ . \tag{11.4}$$

PROOF. We have seen in § 10 that under the assumptions of the theorem, the operator BA is closed. One may consider that BA takes values in the Banach space $R(B)$. By (11.3), its range has a closed direct complement in $R(B)$. This implies (see Theorem 2.4) that $R(BA)$ is closed and that equality (11.4) holds.

The theorem is proven.

THEOREM 11.2. *If the operators* A *and* B *have dense domains in* E *and* F *respectively, and equation* (A) *is* d-*normal, then the operator* BA *has a dense domain in* E.

PROOF. Choose $x \in E$. Given any $\varepsilon > 0$, there is $x_1 \in D(A)$ such that $\|x - x_1\|_E < \varepsilon$. By Lemma 8.1, $D(B) \cap R(A)$ is dense in $R(A)$, and so there is an element $x_2 \in D(A)$ such that

$$Ax_2 \in D(B) \quad \text{and} \quad \|Ax_2 - Ax_1\|_F < \varepsilon \quad .$$

Since (A) is normally solvable, there is an element $\bar{x}$ satisfying

$$A\bar{x} = A(x_2 - x_1)$$

and

$$\|\bar{x}\|_E \leqq k\|A(x_2 - x_1)\|_F \leqq k\varepsilon \quad .$$

If $x_3 = x_1 + \bar{x}$, then

$$\|x - x_3\|_E \leqq \|x - x_1\|_E + \|\bar{x}\|_E \leqq (1 + k)\varepsilon$$

and

$$Ax_3 = Ax_1 + A\bar{x} = Ax_1 + A(x_2 - x_1) = Ax_2 \in \mathcal{D}(B),$$

i.e., $x_3 \in \mathcal{D}(BA)$. Since ε is arbitrary, this completes the proof of the theorem.

OBSERVATION 11.3. Incidentally, we have proved that for each $x_1 \in \mathcal{D}(A)$ one can choose an element $x_3 \in \mathcal{D}(BA)$ such that

$$\| x_3 - x_1 \|_E \leq k\varepsilon \quad \text{and} \quad \| Ax_3 - Ax_1 \|_F = \| Ax_2 - Ax_1 \|_F \leq \varepsilon .$$

In other words, the operator A is the closure of its restriction to $\mathcal{D}(BA)$.

THE ADJOINT EQUATION. Under the assumptions of Theorem 11.2, the operators A^*, B^*, and $(BA)^*$ exist and act from F^* to E^*, from G^* to F^*, and from G^* to E^*, respectively. Moreover, one could consider the operator A^*B^* to be acting also from G^* to E^*. If $h \in \mathcal{D}(A^*B^*)$, then the functional $B^*h \in F^*$ is an element of $\mathcal{D}(A^*)$, i.e.,

$$|A^*B^*h(x)| = |B^*h(Ax)| \leq c\| x \|_E \qquad (x \in \mathcal{D}(A)).$$

If now $x \in \mathcal{D}(BA)$, then $Ax \in \mathcal{D}(B)$ and the last relations imply

$$|A^*B^*h(x)| = |h(BAx)| \leq c\| x \|_E ,$$

giving that $h \in \mathcal{D}((BA)^*)$ and

$$(BA)^*h(x) = h(BAx) \qquad (x \in \mathcal{D}(BA)).$$

Consequently, the linear functionals $A^*B^*h \in E^*$ and $(BA)^*h \in E^*$ are the same on the set $\mathcal{D}(BA)$. Since the latter is dense, these two functionals are equal. In other words, the operator $(BA)^*$ is an extension of the operator A^*B^*.

THEOREM 11.4. *Under the assumptions of Theorem 11.2, operators* A^*B^* *and* $(BA)^*$ *coincide.*

PROOF. Let $h \in \mathcal{D}((BA)^*)$. This means that on $\mathcal{D}(BA)$

$$|(BA)^*h(x)| = | h(BAx)| \leq c\| x\|_E \quad (x \in \mathcal{D}(BA)) \tag{11.5}$$

Since equation (A) is normally solvable, given x one can find $\tilde{x}$ such that $Ax = A\tilde{x}$ and $\| \tilde{x}\|_E \leq k\| Ax\|_F$. Thus

$$|h(BAx)| = |h(BA\tilde{x})| \leq c\| \tilde{x}\|_E \leq ck\| Ax\|_F \quad (x \in \mathcal{D}(BA)). \tag{11.6}$$

Now let $y \in \mathcal{D}(B)$. Using (11.2), we can write $y = Ax + u$, where $x \in \mathcal{D}(BA)$ and $u \in L$. By (11.6),

$$|h(By)| \leq |h(BAx)| + |h(Bu)| \leq ck\| Ax\|_F + \| h\|_{G^*} \| Bu\|_G .$$

Any linear operator on the finite dimensional space L is bounded; hence $\|Bu\|_G \leqq c_1\|u\|_F$, and

$$|h(By)| \leqq c_2(\|Ax\|_F + \|u\|_F) \leqq c_3\|Ax + u\|_F = c_3\|y\|_F .$$

This inequality shows that $h \in \mathcal{D}(B^*)$. Inequality (11.5) can be written in the form

$$|B^*h(Ax)| \leqq c\|x\|_E \quad (x \in \mathcal{D}(BA)).$$

Using Observation 11.3, one may pass to limits and conclude that the last inequality holds for all $x \in \mathcal{D}(A)$. Then $B^*h \in \mathcal{D}(A^*)$ and $h \in \mathcal{D}(A^*B^*)$. Moreover, we have shown that $A^*B^*h = (BA)^*h$, and so complete the proof of the theorem.

§ 12. NOETHERIAN EQUATIONS. INDEX

Equation (A) is called *Noetherian* if it is both n-normal and d-normal; the corresponding operator A is called a *Noetherian* operator or a Φ*-operator*. [In the literature, these operators are sometimes called *Fredholm* operators. The operators A appearing in n-normal or d-normal equations (A) are called *semi-Fredholm*.]

The number

$$\mathrm{ind}(A) = n(A) - d(A)$$

is called the *index of equation* (A) or the *index of the operator* A. [The notion of index is defined even when only one of the numbers $n(A)$ and $d(A)$ is finite, and in this case it can take value $\pm\infty$. Sometimes the index is defined to be the number $-\mathrm{ind}(A)$.]

If $\mathcal{D}(A)$ is dense in E, then the adjoint equation (A^*) of a Noetherian equation (A) is also Noetherian, and

$$\mathrm{ind}(A^*) = n(A^*) - d(A^*) = d(A) - n(A) = -\mathrm{ind}(A).$$

Later on we shall present some rather remarkable properties of the index of a Noetherian equation. Here we discuss only two of them. Let M be an r-dimensional subspace of E such that $M \cap \mathcal{D}(A) = \{\theta\}$, and let $\widetilde{A}$ be some linear extension of the operator A to $\mathcal{D}(A) \oplus M$.

THEOREM 12.1. *If the operator* A *is Noetherian, then the operator* $\widetilde{A}$ *is also Noetherian, and* $\mathrm{ind}(\widetilde{A}) = \mathrm{ind}(A) + r$.

PROOF. The subspace M can be decomposed into a direct sum $M = M_1 \oplus M_2 \oplus M_3$ of three subspaces, where M_1 consists of those elements which are taken by $\widetilde{A}$ into θ, M_2 is mapped one-to-one into $R(A)$ by $\widetilde{A}$, and M_3 is mapped one-to-one into some direct complement L of $R(A)$ $(F = R(A) \oplus L)$. If $\dim M_i = r_i$, $i = 1,2,3$, then it is obvious that the existence of M_1 increases $n(A)$ by r_1, and similarly the existence of M_3 decreases $d(A)$ by r_3.

Let $e_1,\ldots,e_{r_2}$ be a basis for M_2. Since $\widetilde{A}e_k \in R(A)$, there are elements $x_k \in \mathcal{D}(A)$ such that $Ax_k = -\widetilde{A}e_k$, or $\widetilde{A}(x_k + e_k) = \theta$. Thus, the subspace M_2 generates a subspace N_2 of the same dimension, such that $\widetilde{A}x = \theta$ for all $x \in N_2$, and $N_2 \cap \mathcal{D}(A) = \{\theta\}$. Now it is easy to see that the kernel of $\widetilde{A}$ has the form $N(A) \oplus M_1 \oplus N_2$ and has the dimension $n(A) + r_1 + r_2$. Finally

$$\begin{aligned} \mathrm{ind}(\widetilde{A}) &= n(A) + r_1 + r_2 - (d(A) - r_3) = n(A) - d(A) + r = \\ &= \mathrm{ind}(A) + r, \end{aligned}$$

and the theorem is proven.

The results of §§ 10 and 11 show that if A and B are Noetherian and $\overline{\mathcal{D}(B)} = F$, then the operator BA is also Noetherian. Indeed, we have seen in § 10 that the n-normality of A and B implies the n-normality of BA. Moreover, Theorem 11.1 shows that if A is d-normal and B is Noetherian, then BA is d-normal and

$$d(BA) = d(A) - \mathrm{ind}(B) + \dim R(A) \cap N(B). \tag{12.1}$$

Now observe that $R(A) \cap N(B)$ is the image of $N(BA)$ under the map A, whose kernel is $N(A)$. Therefore,

$$\dim R(A) \cap N(B) = n(BA) - n(A). \tag{12.2}$$

If we substitute this into (12.1), we reach the following conclusion.

THEOREM 12.2. *Let* A, B *be Noetherian operators and suppose that* B *has a dense domain. Then*

$$\mathrm{ind}(BA) = \mathrm{ind}(A) + \mathrm{ind}(B).$$

§ 13. EQUATIONS WITH OPERATORS WHICH ACT IN A SINGLE SPACE

Let A be an operator which acts in the space E, has a domain dense in E, and is Noetherian. The operators $A^2, A^3, \ldots$ enjoy these same properties, and this allows us to introduce a new important characteristic of equation (A). Namely, let $N_\infty(A)$ be the set of all elements x such that $A^k x = \theta$ for some $k > 0$. This set is a linear manifold: if $A^{k_1} x_1 = \theta$ and $A^{k_2} x_2 = \theta$, then

$$A^k(x_1 + x_2) = \theta \quad \text{for} \quad k = \max(k_1, k_2).$$

THEOREM 13.1. *If* $N_\infty(A)$ *is finite dimensional, then equation* (A) *has a nonpositive index.*

PROOF. Let $n_\infty = \dim N_\infty(A)$, and assume that $\operatorname{ind}(A) > 0$. By Theorem 12.2, the index of the operator A^k equals $k \operatorname{ind}(A)$, and so $\operatorname{ind}(A^k) = k \operatorname{ind}(A) > n_\infty$ for k large enough. On the other hand, $\operatorname{ind}(A^k) \leq n(A^k) \leq n_\infty$. This contradiction proves the theorem.

COROLLARY. *Assume that the domain of the operator* A* *is also dense. If* $N_\infty(A^*)$ *is finite dimensional, then the index of* (A*) *is also nonpositive, and so the index of* (A) *is nonnegative. If* $N_\infty(A)$ *and* $N_\infty(A^*)$ *are both finite dimensional, then the index of* (A) *is zero.*

A classical example of an equation with index zero is the equation

$$(I + T)x = y, \tag{13.1}$$

where T is a compact operator. Indeed, the adjoint equation

$$(I + T^*)x = y \tag{13.2}$$

has the same form, because the operator T* is also compact. Recalling the last statement of § 6, both equations are n-normal, and as such (see § 8), Noetherian. Let $A = I + T$ and consider the linear manifold $N_\infty(A)$. If $N_k(A)$ denotes the set of all solutions of the equation $A^k x = \theta$, then obviously

$$N_k(A) \subset N_{k+1}(A).$$

If $N_k(A) = N_{k+1}(A)$, then

$$N_m(A) = N_k(A)$$

for all $m > k$. To see this, if we take $x \in N_{k+2}(A)$, then $A^{k+1}(Ax) = \theta$, i.e., $Ax \in N_{k+1}(A)$. This shows that $Ax \in N_k(A)$, and so

$$A^k(Ax) = A^{k+1}x = \theta, \quad \text{i.e.,} \quad x \in N_{k+1}(A),$$

and so on. We see that if $N_\infty(A)$ is infinite dimensional, then every $N_k(A)$ is a proper subspace of $N_{k+1}(A)$. Therefore, we can construct a sequence $x_k \in N_{k+1}(A)$ such that $\| x_k \| = 1$ and

$$\rho(x_k, N_k(A)) \geqq 1/2.$$

Now take $m > k$ and compute

$$\| Tx_m - Tx_k \| = \| Ax_m - Ax_k + x_k - x_m \| .$$

It is clear that $A^m(Ax_m - Ax_k + x_k) = \theta$, i.e., $Ax_m - Ax_k + x_k \in N_m(A)$. Consequently,

$$\| Tx_m - Tx_k \| \geqq \rho(x_m, N_m(A)) \geqq 1/2.$$

In other words, the sequence $\{Tx_k\}$ is not compact, which contradicts the assumption that T is a compact operator. We conclude that $\dim N_\infty(A) < \infty$.

Similarly, $\dim N_\infty(I + T^*) < \infty$.

An equation of the form (13.1) will be called a *canonical Fredholm equation*. The theorem we proved can be restated as follows.

THEOREM 13.2. *A canonical Fredholm equation has index zero.*

§14. FREDHOLM EQUATIONS. REGULARIZATION OF EQUATIONS

Equation (A) is called a *Fredholm equation* if it is Noetherian and has index zero. The corresponding operator A is called a *Fredholm operator*.

If B is also a Fredholm operator with dense domain, then by Theorem 12.2, the operator BA is Fredholm.

Each operator A having a bounded inverse $U = A^{-1}$ defined on the entire space F is obviously Fredholm $(n(A) = d(A) = ind(A) = 0)$.

Consider an operator A of the form

$$A = U^{-1} + T , \tag{14.1}$$

where U is a bounded invertible operator from F to E, and T is a compact operator from E to F. Then $A = (I + TU)U^{-1}$. The operator TU is compact in F, and so $I + TU$ is a canonical Fredholm operator. Therefore, A, as a product of two Fredholm operators, is also Fredholm.

If we transform equation (A) using the operator U, then it becomes a canonical Fredholm equation

$$(I + UT)x = Uy \tag{14.2}$$

in the space E. Moreover, since U is bounded and invertible, (14.2) is equivalent to (A).

If equation (A) can be reduced to a canonical Fredholm one by means of a bounded transformation B, then one says that (A) admits a *left regularizer* B. If, in addition, the equation $BAx = By$ is equivalent to equation (A), then one says that B is an *equivalence-regularizer*.

In the example discussed above, U is a left equivalence-regularizer.

Now let A be an arbitrary Fredholm operator. Then the spaces $N(A)$ and $R^{\perp}(A)$ are finite dimensional and have the same dimension. Consequently, there are decompositions $E = N(A) \oplus E_1$, $F = R(A) \oplus L$, where E_1 is closed and $\dim L = n(A)$. Moreover, there is an operator C which takes L one-to-one onto $N(A)$, and we satisfy the conditions considered at the end of § 10. Recall that the transformation operator B was defined there to be A_1^{-1} on $R(A)$ (where A_1 denotes the restriction of A to E_1), to be C on L, and then extended by linearity to F. As a result, B generated a bounded equivalence such that

$$BAx = (I - P)x , \tag{14.3}$$

where P is the projection onto the subspace $N(A)$. In our case, P is a finite rank operator, and hence, by (14.3),

$$A = B^{-1} + T,$$

with $T = -B^{-1}P = -C^{-1}P$ of finite rank.

In other words, we get

THEOREM 14.1. *Every operator of the form* (14.1) *, where* U *is a bounded operator from* F *to* E*, and* T *is a compact operator from* E *to* F*, is Fredholm. Any Fredholm operator has a representation of the form* (14.1)*, where* T *is a finite rank operator. Every Fredholm equation admits a left equivalence-regularizer.*

Now let (A) be a Noetherian equation of index ind(A). One can always construct a bounded Noetherian operator B_1 mapping F into some other space G and having a given index ind(B) = -ind(A). Indeed, if $\mathrm{ind}(A) \geqq 0$, then one can take $G = F \times M$, where M is a space with $\dim M = \mathrm{ind}(A)$, and set $B_1y = (y,\theta)$ $(y \in F)$. We have $n(B_1) = 0$, $d(B_1) = \mathrm{ind}(A)$, and $\mathrm{ind}(B_1) = -\mathrm{ind}(A)$. If $\mathrm{ind}(A) < 0$, then decompose F into a direct sum $F = R(A) \oplus L$, and choose a subspace L_1 in L with $\dim L_1 = n(A) < d(A) = \dim L$. Further, set $G = R(A) \oplus L_1$ and choose some projection B_1 of F onto G. Then

$$d(B_1) = 0, \quad n(B_1) = \dim L - \dim L_1 = d(A) - n(A) = -\mathrm{ind}(A).$$

By theorem 12.2, B_1A has index zero and is a Fredholm operator. Moreover, Theorem 14.1 shows that B_1A can be represented as $B_1A = U^{-1} + T$.

This shows that the operator $B = UB_1$ is a left regularizer for equation (A):

$$BAx = UB_1Ax = U(U^{-1} + T)x = (I + UT)x \quad (x \in D(A)). \tag{14.4}$$

We obtain the following statement.

THEOREM 14.2. *Every Noetherian equation has a left regularizer. If the index of the equation is nonnegative, then there is a left equivalence-regularizer.*

The last conclusion is a result of the construction of B_1.

Theorem 14.2 must be applied with some caution. The problem is that equality (14.4) is valid on $\mathcal{D}(A)$, while the operator $I + UT$ can be extended by continuity to the entire $\overline{\mathcal{D}(A)}$. If we do extend it, the defect of its range does not change, in constrast to the dimension of its kernel which can do so. Furthermore, in applications, the operator $I + UT$ often has a natural extension to the entire space E, and for this particular extension both the numbers discussed above may alter. Consequently, the theorems concerning the relations between **the** numbers $n(A)$, $d(A)$, $ind(A)$ and the corresponding numbers for the resulting canonical Fredholm equation should be used only while taking the above circumstances into consideration. If operator A is defined on the entire space E, then no special precautions are necessary.

We remark that an equation with a closed operator A which admits a left regularizer is n-normal. Indeed, in this case BA has the form $I + T$, where T is compact on $\mathcal{D}(A)$. Then the operator $\overline{BA}$ is n-normal, which, as was shown in § 10, implies that equation (A) is n-normal. Conversely, let the equation (A) be n-normal. If (A) is not Noetherian, then in the case $ind(A) < 0$, one can repeat the arguments appearing in the proof of Theorem 14.2 in order to construct the operator B_1, assuming that $R(A)$ has a closed direct complement in F. At the same time, B_1A will be again a Fredholm operator, and we can construct a left regularizer. Therefore, we get

THEOREM 14.3. *If the operator* A *is closed, then a sufficient condition for equation* (A) *to be* n-*normal is the existence of a left regularizer. When* $R(A)$ *has a closed direct complement in* F, *this condition is also necessary.*

Exhibiting a left regularizer is one of the most important ways of proving that a given equation is n-normal.

Now we would like to settle the question of the existence of an equivalence-regularizer more precisely. Again here it is essential to distinguish between the cases where the operator A is defined on the entire space and bounded and where this is not so. In the first case, a left equivalence-regularizer B leads to the equality $BA = I + T$ on the entire space E. This means that B maps $R(A)$ onto $R(I + T)$, and hence that B is d-normal. Moreover, $N(B) = \{\theta\}$, and thus B is Noetherian. We also get a decomposition

$$R(B) = R(BA) \oplus M_1 = R(I + T) \oplus M_1,$$

where M_1 is some subspace of a finite dimensional complement of the range $R(I + T)$ in E. Since B is defined on the entire space F and B is invertible, this decomposition implies that $F = R(A) \oplus L$, where L is the preimage of M_1 under B.

Therefore, def $\overline{R(A)} < \infty$, and using Theorem 14.3, A is Noetherian.

Finally,

$$\text{ind}(A) + \text{ind}(B) = \text{ind}(I + T) = 0,$$

and so

$$\text{ind}(A) = -\text{ind}(B) = d(B) \geqq 0.$$

We obtain

THEOREM 14.4. *Let the operator* A *be defined on the entire space* E *and be bounded. Then there is a left equivalence-regularizer for equation* (A) *if and only if* (A) *is Noetherian and has nonnegative index.*

When an n-normal operator A is not bounded, its domain is not closed. Construct a decomposition $E = N(A) + E_1$. Then $\mathcal{D}_1 = \mathcal{D}(A) \cap E_1$ does not coincide with E_1, and so we can choose an algebraic direct complement S of $\mathcal{D}_1$ in E_1: $E_1 = S + \mathcal{D}_1$. If $\dim S \geqq d(A)$ and $d(A) < \infty$, then one can find a linear operator C mapping L one-to-one into S, where $F = R(A) \oplus L$. Now set B equal to A_1^{-1} on $R(A)$ (where A_1 is the restriction of A to E_1) and equal to C on L. The resulting linear operator B is a left equivalence-regularizer transforming (A) to the form

$$BAx = (I - P)x = By,$$

where P is the finite rank projection onto $N(A)$.

In other words, if $\dim S \geqq d(A) < \infty$, then there exists a left equivalence-regularizer.

It is obvious that there are situations when $\dim L = \infty$ and $\dim S = \infty$, and one can still construct on L a bounded linear operator C which maps L one-to-one onto its image in S, and thus defines a left equivalence-regularizer for equation (A). Therefore, *the existence of a left equivalence-regularizer for an equation with an unbounded operator does not necessarily imply that the given equation is Noetherian* (in the book [57], a statement concerning this point is not exact).

§15. LINEAR CHANGES OF VARIABLE

In equation (A) one can make a change of the unknown element by taking $x = Cs$, where C is some linear operator from a Banach space H into the Banach space E. For the new unknown element, we obtain the equation

$$ACs = y . \tag{AC}$$

Any solution of equation (AC) produces a solution $x = Cs$ of equation (A). However, one can loose some of the solutions of (A), namely those which cannot be expressed as Cs. To eliminate such a possibility, it is necessary and sufficient to assume that $\mathcal{D}(A) \subset \mathcal{R}(C)$. Whenever this condition is fulfilled, we shall say that the change of variable C is an *equivalence*.

If C is an equivalence, then $\mathcal{R}(A) = \mathcal{R}(AC)$, and so the equations (A) and (AC) are, or are not, simultaneously normally solvable, d-normal, densely, or everywhere solvable. Furthermore, $d(A) = d(AC)$, and the dimension of the kernel does not decrease, i.e., $n(A) \leq n(AC)$; consequently, $\operatorname{ind}(A) \leq \operatorname{ind}(AC)$. [Here we allow the index to take infinite values.] Finally, if $N(C) = \{\theta\}$, then $n(A) = n(AC)$ and $\operatorname{ind}(A) = \operatorname{ind}(AC)$, and we state the following useful result.

THEOREM 15.1. *If* C *is an equivalence such that equation* (AC) *is Noetherian, then the equation* (A) *is also Noetherian and* $\operatorname{ind}(A) \leq$ $\leq \operatorname{ind}(AC)$. *If in addition* C *is one-to-one, then* $\operatorname{ind}(A) = \operatorname{ind}(AC)$.

If C is not an equivalence, then $\mathcal{R}(A) \supset \mathcal{R}(AC)$. If equation (AC) is d-normal, then $\mathcal{R}(AC)$ is closed and $\mathcal{R}(A)$ differs from $\mathcal{R}(AC)$ by a finite dimensional direct complement, and hence $\mathcal{R}(A)$ is also closed. Moreover, $d(A) \leq d(AC)$. Therefore, *if equation* (AC) *is* d-*normal for a change of variable* C, *then equation* (A) *is also* d-*normal, and* $d(A) \leq d(AC)$.

A bounded operator C (acting from F to E) is called a *right regularizer* for equation (A) if the linear change of variable $x = Cs$ reduces (A) to a canonical Fredholm equation (AC) in F. If the change of variable $x = Cs$ is also an equivalence, then C is called a *right equivalence-regularizer*.

The discussion above shows that the existence of a right regularizer implies the d-normality of the given equation (A). Furthermore, if this regularizer is an equivalence, then equation (A) is Noetherian and its index is at most the index of equation (AC), which is zero: $\mathrm{ind}(A) \leq 0$. Conversely, let (A) be Noetherian. If $\mathrm{ind}(A) \leq 0$, then consider the new space $H = E \times R$, where R is a space of dimension $-\mathrm{ind}(A)$, and set $C_1(x,w) = x$. Then the operator C_1 acts from H to E and the new equation (AC_1) has index zero. Consequently,

$$AC_1 = U^{-1} + T_1,$$

where U is a bounded operator from the entire space F to H, and T_1 is a compact operator from H to F.

In other words, the change of variable $x = C_1 Us = Cs$ reduces equation (A) to the canonical Fredholm form

$$ACs = (I + T_1 U)s = y .$$

Since C_1 maps the entire space H onto E and U maps the entire space F onto the domain of the operator AC_1, the change of variable $x = Cs$ is an equivalence.

Now assume that $\mathrm{ind}(A) > 0$ and that the kernel $N(A)$ has a closed direct complement E_1. Choose some subspace N_1 in $N(A)$ with $\dim N_1 = d(A)$, and denote by H the subspace $E_1 \oplus N_1$ of E. Then the identity operator C_1, considered as acting from H into E, reduces (A) to a Fredholm equation. Repeating the previous arguments, one can construct a right regularizer for (A). However, this time the regularizer is not an equivalence.

We have proved the following theorem.

THEOREM 15.2. *If the operator* A *is closed, then the existence of a right regularizer is sufficient for equation* (A) *to be* d-*normal. If* $N(A)$ *has a closed direct complement, then this condition is also necessary. There exists a right equivalence-regularizer if and only if equation* (A) *is Noetherian and has nonpositive index.*

OBSERVATION 15.3. If A is closed and C is bounded, then the new operator AC is also closed, as one can easily check. Due to this fact, the dangers, discussed in § 14, surfacing during the left regularization of an equation are now absent.

Among the main examples of equations which can be regularized are the equations of the type

$$(I - S)x = y , \tag{15.1}$$

where S is a bounded operator acting in space E and such that some power $S^m = T$ of S is a compact operator. Both a left and right regularizer for equation (15.1) is given by

$$B = I + S + \dots + S^{m-1}.$$

Indeed, $B(I - S) = (I - S)B = I - T$. Equation (15.1) is therefore Noetherian. However, the operator B can have a nontrivial kernel $N(B)$, and hence no statement concerning the index of (15.1) can be made at the moment. Using properties of compact operators, one can show that there are arbitrarily large numbers n such that the kernels $N(I - \varepsilon_i S)$ of the operators $I - \varepsilon_i S$ are trivial for all the n-th roots $\varepsilon_i \neq 1$ $(i = 1,\dots,n-1)$ of the identity. Then the operator

$$C = \prod_{i=1}^{n-1}(I - \varepsilon_i S) = I + S + \dots + S^{n-1}$$

defines a one-to-one mapping of E onto itself, and the change of variable $x = Cs$ reduces (15.1) to the form $(I - S)Cs = (I - S^n)s = y$. If we choose $n \geqq m$, then S^n is compact and the resulting equation is Fredholm. Therefore, Theorem 15.1 implies

THEOREM 15.4. *Suppose that equation* (15.1) *is given by a bounded operator* S *which acts in* E *such that one of the powers of* S *is compact. Then* (15.1) *is a Fredholm equation.*

§ 16. STABILITY OF THE PROPERTIES OF AN EQUATION

In this section we consider, along with equation (A), the perturbed equation

$$Ax + Qx = y \ , \quad (x \in \mathcal{D}(A)) \ . \qquad (A + Q)$$

If a certain property of equation (A) is preserved as we pass to equation (A + Q) for all operators Q in a certain class, then we say that this property is *stable to perturbations from the given class.*

To begin with, take the perturbations to be bounded operators defined on the entire space E and having small enough norms ("small perturbations").

If equation (A) *is correctly solvable, then for small* $\|Q\|_{E \longrightarrow F}$ *equation* (A + Q) *is also correctly solvable.*

Indeed,

$$\|x\|_E \leqq k\|Ax\|_F \leqq k\|(A + Q)x\|_F + k\|Q\|_{E \longrightarrow F}\|x\|_E$$

for all $x \in \mathcal{D}(A)$. If $k\|Q\|_{E \longrightarrow F} \leqq 1$, then

$$\|x\|_E \leqq k_1\|(A + Q)x\|_F \quad (k_1 = k(1 - k\|Q\|_{E \longrightarrow F})^{-1}),$$

i.e., equation (A + Q) is correctly solvable.

The property of being uniquely solvable is not stable if the original equation is not correctly solvable.

Namely, if equation (A) is not correctly solvable, then there is a sequence $\{x_n\}$ such that $\|x_n\| = 1$ and $\|Ax_n\| \longrightarrow 0$. Choose functionals f_n satisfying $f_n(x_n) = 1$, $\|f_n\| = 1$, and define the operators Q_n by $Q_n(x) = -f_n(x)Ax_n$. We get

$$\|Q_n\| = \|Ax_n\| \longrightarrow 0 \quad \text{and} \quad (A + Q_n)x_n = Ax_n - f_n(x_n)Ax_n = \theta,$$

i.e., equation (A + Q) is not uniquely solvable.

If the operator A *is closed, then so is the operator* $A + Q$, *where* Q *is any bounded operator.*

Indeed, let $x_n \longrightarrow x$, $(A + Q)x_n \longrightarrow y$. Then $Qx_n \longrightarrow Qx$ (because Q is bounded), and so $Ax_n \longrightarrow y - Qx$. Since A is closed,

$x \in \mathcal{D}(A) = \mathcal{D}(A + Q)$ and $Ax = y - Qx$, i.e., $(A + Q)x = y$.

Let A *be closed. For equation* (A) *the property of dense solvability is not stable if* (A) *is not ewerywhere solvable.*

Assume that $\overline{\mathcal{D}(A)} = E$. If equation (A) is densely but not everywhere solvable, then it is not normally solvable. By Theorem 5.1, equation (A^*) is not closedly solvable, and so, certainly, not correctly solvable. Therefore, one can find a sequence of functionals $g_n \in F^*$, such that $\| g_n \| = 1$ and $A^*g_n \longrightarrow 0$. Choose, for each g_n, an element $y_n \in F$ satisfying $\| y_n \| = 1$ and $g_n(y_n) \geq 1/2$, and define the operators Q_n by

$$Q_n(x) = -(A^*g_n)(x) \frac{y_n}{g_n(y_n)} .$$

We have

$$\| Q_n \| = \| A^*g_n \| \frac{\| y_n \|}{|g_n(y_n)|} \leq 2 \| A^*g_n \| \longrightarrow 0 .$$

Further,

$$g_n((A + Q_n)x) = g_n(Ax) - (A^*g_n)(x) \frac{g_n(y_n)}{g_n(y_n)} = 0$$

for all $x \in \mathcal{D}(A)$. Consequently, $g_n \perp \mathcal{R}(A + Q_n)$, i.e., equation $(A + Q_n)$ is not densely solvable.

If the domain $\mathcal{D}(A)$ is not dense in E, then A can be interpreted as an operator acting from $E' = \overline{\mathcal{D}(A)}$ into F. Then the functionals A^*g_n are elements of E'^*, and so are defined only on E'. Extending them, while preserving the norms, to functionals f_n defined on the entire space E, and defining operators Q_n by

$$Q_n(x) = -f_n(x) \frac{y_n}{g_n(y_n)} ,$$

we reach the same conclusion.

We emphasize that, in fact, we have shown that, for an equation which is not correctly solvable, the properties of unique and dense solvability are not stable even to rank one perturbations.

If equation (A) *is* n-*normal, then for* $\| Q \|_{E \longrightarrow F}$ *small enough equation* $(A + Q)$ *is also* n-*normal, and* $n(A) \geq n(A + Q)$.

Decompose E as $E = N(A) \oplus E_1$ and denote by A_1 and Q_1 the restrictions of A and Q to E_1. The operator A_1 maps $\mathcal{D}_1 = \mathcal{D}(A) \cap E_1$ onto $\mathcal{R}(A)$ and has a bounded inverse, i.e.,

$$\| x \|_E \leq k \| A_1 x \|_F \quad (x \in \mathcal{D}_1).$$

Then for $k\|Q\|_{E\to F} < 1$, equation

$$(A_1 + Q_1)x = y \qquad (A_1 + Q_1)$$

is correctly solvable (i.e., the operator $A_1 + Q_1$ has a bounded inverse), and thus normally solvable. The operator $A + Q$ is obtained from $A_1 + Q_1$ by an extension of finite dimension $n(A)$. Under such an extension normal solvability is preserved, while the dimension of the kernel does not increase by more than $n(A)$ (see the proof of Theorem 12.1). Therefore, both our claim and the inequality $n(A) \geqq n(A + Q)$ have been proven.

OBSERVATION 16.1. Assume that the kernel $N(A)$ of the closed operator A has a closed direct complement E_1 in E. Then, using the previous proof, when $\|Q\|_{E\to F}$ is small enough, the kernel $N(A + Q)$ of $A + Q$ intersects the subspace E_1 only at zero.

If equation (A) *is Noetherian, then for* $\|Q\|_{E\to F}$ *small enough equation* (A + Q) *is also Noetherian, and* $\operatorname{ind}(A) = \operatorname{ind}(A + Q)$.

Consider the restrictions A_1, Q_1 of the operators A, Q to the subspace E_1. On this subspace one can rewrite equation (A + Q) as

$$(A + Q)x = (A_1 + Q_1)x = (I + Q\tilde{A}_1^{-1})A_1x = y \quad (x \in E_1) \qquad (16.1)$$

where $\tilde{A}_1^{-1}$ is an extension of the operator A_1^{-1} to a bounded operator defined on the entire space F. By theorem 12.1, $\operatorname{ind}(A_1) = -d(A)$. If $\|Q\|_{E\to F}\|\tilde{A}_1^{-1}\|_{F\to E} < 1$, then the operator inside the brakets has a bounded inverse in F, and so it is Fredholm. By Theorem 12.2, equation $(A_1 + Q_1)$ is Noetherian, and $\operatorname{ind}(A_1 + Q_1) = -d(A)$. The operator $A + Q$ can be interpreted as an extension of $A_1 + Q_1$ by a finite dimensional subspace $N(A)$. Finally, by Theorem 12.1,

$$\operatorname{ind}(A + Q) = \operatorname{ind}(A_1 + Q_1) + n(A) = n(A) - d(A) = \operatorname{ind}(A).$$

Let us write down our conclusions.

THEOREM 16.2. *For any equation, the properties of being correctly solvable,* n-*normal,* d-*normal, or Noetherian are stable to small perturbations. For small enough perturbations, the index of a Noetherian equation does not change.*

For perturbations of a more general form, the stability of d-normality will be proven immediately below.

An interesting problem is that of the stability of normal solvability when $n(A) = d(A) = \infty$. However, there are equations with closed operators and which satisfy $n(A) = d(A) = \infty$, for which normal solvability is not a stable property.

To see this, assume for a start that $N(A)$ and $R(A)$ have closed direct complements E_1 and L in the spaces E and F respectively. Since $N(A)$ and L are infinite dimensional, one can exhibit (see below) a bounded operator Q from $N(A)$ to L whose range $R(Q)$ is not closed in L. Now extend Q to a bounded linear operator on E, setting $Qx = \theta$ for all $x \in E_1$. We claim that the range $R(A + \varepsilon Q)$ is not closed for arbitrary $\varepsilon \neq 0$. Indeed, let $y_0 \in \overline{R(Q)}$, $y_0 \notin R(Q)$. Then there is a sequence $x_n \in N(A)$ such that $Qx_n \longrightarrow y_0$. Since

$$(A + \varepsilon Q)(\frac{1}{\varepsilon} x_n) = Qx_n \longrightarrow y_0,$$

$y_0 \in \overline{R(A + \varepsilon Q)}$. However, $y_0 \notin R(A + \varepsilon Q)$, since, by assuming the contrary, we obtain

$$y_0 = Ax_0 + \varepsilon Qx_0 \quad (x_0 \in \mathcal{D}(A)),$$

or

$$y_0 - \varepsilon Qx_0 = Ax_0$$

which, using the direct sum decomposition $F = R(A) \oplus L$, shows that $y_0 = \varepsilon Qx_0$, i.e., $y_0 \in R(Q)$; contradiction.

To construct Q as promised, choose in $N(A)$ an infinite minimal sequence of elements x_n (minimal meaning that no element belongs to the closed linear span of the other elements), and choose a sequence of linear functionals f_n, biorthogonal to x_n ($f_k(x_n) = \delta_{kn}$, the Kronecker symbol). Now define Q on the entire space E by the formula

$$Q(x) = \sum_{k=1}^{\infty} \lambda_k f_k(x) y_k,$$

where $\sum|\lambda_k| < \infty$, and $\{y_k\}$ is some sequence of linearly independent elements of L satisfying $\|y_k\| = 1$. Then Q, as an operator from $N(A)$ into L, is compact and of infinite rank ($Q(x_n) = \lambda_n y_n$). Consequently, its range is not closed (see p.7). We remark that here we did not use the fact that $N(A)$ has a closed direct complement in in order to construct the operator Q, and hence one can drop this assumption. Similarly, the assumption that $R(A)$ has a closed direct complement in F is not necessary. Indeed, the only fact we needed to get Q was that one can find an infinite dimensional linear manifold L in F such that $L \cap R(A) = \{\theta\}$. It can be shown that for def $R(A) < \infty$ such a linear manifold L always exists (see [29]).

Therefore, if one should decide to study only those properties of an equation which are stable to small perturbations, one would have to restrict oneself to the properties listed in Theorem 16.2.

Now consider perturbations of equation (A) by unbounded operators. Assume that the operator A is closed, while the operator Q is defined on $\mathcal{D}(A)$ and can be extended to a closed operator $\overline{Q}$. As was shown in § 2, $\mathcal{D}(A)$ can be made into a Banach space E_A with norm $\|x\|_{E_A} = \|x\|_E + \|Ax\|_F$. If we consider Q as an operator from E_A into F, then it is closed, and hence bounded. Indeed, let $x_n \longrightarrow x$ in E_A and $Qx_n \longrightarrow y$. Then $x_n \longrightarrow x$ in E and since $\overline{Q}$ is closed, we get $x \in \mathcal{D}(\overline{Q})$ and $\overline{Q}x = y$. But $x \in E_A = \mathcal{D}(A)$ and $\overline{Q}$ equals Q on $\mathcal{D}(A)$, which implies $Qx = y$. Thus Q is bounded. If $\|Q\|_{E_A \longrightarrow F}$ is small enough, then we shall say that Q is a *relatively small perturbation*. (We remark that for a bounded operator Q, $\|Q\|_{E_A \longrightarrow F} \leq \|Q\|_{E \longrightarrow F}$).

For equations (A) and (A + Q) the properties of being uniquely or densely solvable, correctly solvable, n-normal, d-normal, or Noetherian do not depend upon the fact that we regard A and $A + Q$ as operators from E to F or as operators from E_A to F. Therefore, the previous results remain valid for relatively small perturbations.

Let us show that d-normality is stable when A and Q are bounded operators defined on the entire space E. Under this assumption on A and Q, the adjoint equation (A^*) is n-normal. If $\|Q\|_{E \longrightarrow F}$ is small, then so is $\|Q^*\|_{F^* \longrightarrow E^*}$, which implies that equation $(A^* + Q^*)$ is n-normal and $n(A^*) \geq n(A^* + Q^*)$. We conclude that equation (A + Q) is d-normal and $d(A) \geq d(A + Q)$. If we now take E_A instead of E, we see that d-normality is stable to relatively small perturbations.

THEOREM 16.3. *For an equation, the properties of being correctly solvable,* n*-normal,* d*-normal, or Noetherian are stable to relatively small perturbations. If* $\|Q\|_{E_A \longrightarrow F}$ *is small enough and equation* (A) *is Noetherian, then* $\mathrm{ind}(A + Q) = \mathrm{ind}(A)$.

Finally, let us consider compact or A-compact (i.e., compact as operators from E_A into F) perturbations. Here we must say from the beginning that correct solvability is not a stable property.

Indeed, let $x_0 \in \mathcal{D}(A)$ and let $f \in E^*$ be a functional satisfying $f(x_0) = 1$. Construct the rank one operator $Q(x) = -f(x)Ax_0$. Then $(A + Q)x_0 = \theta$, and equation (A + Q) is not uniquely, and so not correctly solvable.

We have shown in §6 that n-normality is stable to compact perturbations. By passing to the adjoint equation, one can show that d-normality is also stable to such perturbations.

Finally, to show that the property of being Noetherian is also stable, we appeal again to the representation (16.1). Now $I + Q\tilde{A}_1^{-1}$ is a canonical Fredholm operator, and using again Theorems 12.1 and 12.2, we see that equation $(A + Q)$ is Noetherian and that $\text{ind}(A + Q) = \text{ind}(A)$.

THEOREM 16.4. *The properties of being* n-*normal,* d-*normal, or Noetherian are stable to arbitrary compact or* A-*compact perturbations. Under such perturbations, the index does not change* .

§ 17. OVERDETERMINED EQUATIONS

Sometimes we have to deal with equations (A) such that one can see from their structure immediately that they cannot be solved for all $y \in F$. For example, this happens if there is a closed operator Φ, acting from F into another Banach space G, and such that $\Phi A = 0$. Then equation $Ax = y$ can be solved only for $y \in N(\Phi)$. Relative to a given pair of spaces E and F, an equation which can be solved only when the right-hand side is contained in a subspace F_1 of F, must be naturally considered as *overdetermined*.

Thus, let A be a closed operator such that $\mathcal{D}(A)$ is dense in E and $R(A) \subset F_1$, where F_1 is a subspace of F. Then, of course, one can forget about the space F, and, for example, one can restate all the criteria listed in § 5 for various form of solvability, this time for equation (A) in the subspace F_1. This would involve various properties of equation (A^1*), where A^1* is the adjoint of A when A is considered as an operator from E to F_1. However, more often it is convenient to state such criteria in terms of the properties of the adjoint A^* of operator A, acting from E into F. This can be done, for example, when F_1 has a closed direct complement F_2 in F: $F = F_1 \oplus F_2$. By the Hahn-Banach theorem, any functional $\tilde{g} \in F_1^*$ is the restriction to F_1 of some functionals belonging to F^*. Among all these functionals, choose that one defined by

$$g^0(y) = \begin{cases} \tilde{g}(y) & y \in F_1, \\ 0 & y \in F_2. \end{cases} \tag{17.1}$$

The functional g^0 is in F^*, since for $y = y_1 + y_2$ $(y_i \in F_i)$

$$|g^0(y)| = |\tilde{g}(y_1)| \le \|\tilde{g}\|_{F_1^*} \|y_1\| \le \|\tilde{g}\|_{F_1^*} (\|y_1\| + \|y_2\|) \le$$
$$\le c\|\tilde{g}\|_{F_1^*} \|y\| .$$

Obviously, $\|g^0\|_{F^*} \ge \|\tilde{g}\|_{F_1^*}$, while the previous inequality shows that $\|g^0\|_{F^*} \le c\|\tilde{g}\|_{F_1^*}$. Therefore, (17.1) establishes a one-to-one and bicontinuous correspondence between the space F_1^* and the subspace $F_2^{\perp}$ of F^* consisting of all functionals of F^* which are zero on F_2.

The operator A^{1*} is defined on all functionals $\tilde{g} \in F_1^*$ such that $|\tilde{g}(Ax)| \leq c\|x\|$ $(x \in \mathcal{D}(A))$. Since $R(A) \subset F_1$, we have $\tilde{g}(Ax) = g^0(Ax)$, and hence $g^0 \in \mathcal{D}(A^*)$. Furthermore

$$A^{1*}\tilde{g}(x) = \tilde{g}(Ax) = g^0(Ax) = (A^*g^0)(x) \quad (x \in \mathcal{D}(A)),$$

and since $\mathcal{D}(A)$ is dense in E, $A^{1*}\tilde{g}$ and A^*g^0 coincide as functionals of E^*. Reversing the argument, we see that if $g^0 \in \mathcal{D}(A^*) \cap F_2^{\perp}$, then the restriction $\tilde{g}$ of g^0 to F_1 is an element of $\mathcal{D}(A^{1*})$, and $A^{1*}\tilde{g} = A^*g^0$. In other words, the operator A^{1*} is the restriction of the operator A^* to $\mathcal{D}(A^*) \cap F_2^{\perp}$.

Now we can reformulate the results of §§5 and 8 as follows:

Let A *be a closed operator with a dense domain in* E *and* $R(A) \subset F_1$, *where* F_1 *is a subspace of* F *having a closed direct complement* F_2.

1) *Equation* (A) *is densely solvable in the subspace* F_1 *if and only if* $N(A^*) \cap F_2^{\perp} = \{\theta\}$.

2) *Equation* (A) *is everywhere solvable in the subspace* F_1 *if and only if inequality*

$$\|g\|_{F^*} \leq k\|A^*g\|_{E^*}$$

holds for every $g \in \mathcal{D}(A^*) \cap F_2^{\perp}$.

3) *Equation* (A) *is correctly solvable if and only if equation* (A*) *has, for any* $f \in E^*$, *a solution which belongs to* $F_2^{\perp}$.

4) *Equation* (A) *is* d-*normal relative to the subspace* F_1 *if and only if* $R(A^*)$ *is closed and the intersection* $N(A^*) \cap F_2^{\perp}$ *is finite dimensional.*

When F is a Hilbert space, one can take F_2 to be the orthogonal complement to F_1, and then $F_2^{\perp} = F_1$. Consequently, the criteria listed above become easy to use and verify.

There is another method which is used sometimes to study overdetermined equations. Namely, one constructs a linear operator B acting from some Banach space E_2 into space F_2, and then one considers the equation

$$Ax + Bu = y \quad (x \in \mathcal{D}(A) \subset E, \quad u \in \mathcal{D}(B) \subset E_2). \tag{A,B}$$

Obviously, when we try to solve this new equation , there are no a priori restrictions on the right-hand side. We should mention that the solvability of equation (A,B) for $y \in F_1$ implies that $Bu = y - Ax \in$

$\in F_1$. Since $F_1 \cap F_2 = \{\theta\}$, $Bu = \theta$. Thus, given any $y \in F_1$, any solution (x,u) of equation (A,B) gives a solution x of equation (A). Conversely, any solution x of (A) yields the solution (x,θ) of (A,B). From this we can deduce the following properties.

If equation (A,B), *having an operator acting from* $E \times E_2$ *into* F, *is normally solvable, correctly solvable, or everywhere solvable,* n-*normal,* d-*normal, or Noetherian, then equation* (A) *is respectively normally solvable, correctly solvable, or everywhere solvable in the subspace* F_1, n-*normal,* d-*normal, or Noetherian relative to the subspace* F_1.

The only part that remains to be checked is that the d-normality of equation (A) follows from the d-normality of equation (A,B). This is done using the same arguments invoked when we analysed the connection between the d-normality of equation (A*) and the d-normality of equation ($\hat{A}$*) (see p.26).

§ 18. UNDETERMINED EQUATIONS

Consider the equation

$$Ax + Bu = y \quad (x \in \mathcal{D}(A),\ \ u \in \mathcal{D}(B),\ \ y \in F), \tag{A,B}$$

where A and B are linear operators acting from Banach spaces E and G, respectively, into the Banach space F. If $R(A) \cap R(B) \neq \{\theta\}$, then equation (A,B) is manifestly not uniquely solvable. Indeed, let $Ax_0 = Bu_0$ $(x_0 \in \mathcal{D}(A),\ u_0 \in \mathcal{D}(B))$. Then the pair $(x_0,-u_0)$ is a solution of the homogeneous equation (A,B). Due to such behaviour, it is natural to say that equation (A,B) is *undetermined*.

Certainly, equation (A,B) could be looked upon as an ordinary equation, given by a linear operator $\mathcal{A}$ acting from the space $E \times G$ into the space F by $\mathcal{A}(x,u) = Ax + Bu$. Then the various solvability properties of equation (A,B) can be formulated in terms of the properties of the operator $\mathcal{A}$. However, in applications it is convenient to formulate these solvability conditions in terms of the properties of the operators A and B separately. Often, in doing so, one of the operators, for example A, plays the basic role, while the second operator, for example B, plays an auxiliary role.

Obviously, the set of right-hand sides for which equation (A,B) is solvable equals $R(A) + R(B)$. Any functional from F^*, which is zero on both $R(A)$ and $R(B)$, is zero on $R(A) + R(B)$, and conversely. This shows that *equation* (A,B) *is densely solvable if and only if* $R(A)^{\perp} \cap R(B)^{\perp} = \{\theta\}$.

If $\mathcal{D}(A)$ and $\mathcal{D}(B)$ are dense in E and G, respectively, then the last condition is equivalent to

$$N(A^*) \cap N(B^*) = \{\theta\}.$$

From now on, assume that equation (A) is normally solvable and that the range $R(A)$ has in F a closed direct complement L: $F = R(A) \oplus L$. Denote by P the projection onto $R(A)$ determined by this decomposition. Then equation (A,B) can be written as the equivalent system of equations

$$Ax + PBu = Py,$$
$$(I - P)Bu = (I - P)y.$$

Given some $y \in F$, if equation (A,B) is solvable, then the system is also solvable for this y, and, in particular, so is its second equation. But the converse statement is the important result.

If the second equation of the system is solvable for an element $y \in F$, *then equation* (A,B) *is also solvable for the same* y.

To see this, let u be a solution of the second equation. Then $Py - PBu \in R(A)$, and hence there exists $x \in \mathcal{D}(A)$ such that $Ax = Py - PBu$. Now the pair (x,u) is a solution of equation (A,B). Due to this, we say that the second equation of the system is the *determinative equation* for equation (A,B).

The operator $(I - P)B$ in the determinative equation acts from G into L. Now the set of right-hand sides $y \in F$ for which equation (A,B) is solvable is clearly equal to $R(A) \oplus R((I - P)B)$. We obtain

THEOREM 18.1. *If equation* (A) *is normally solvable and* $R(A)$ *has a closed direct complement in* F, *then equation* (A,B) *is densely solvable, normally solvable,* d-*normal, or everywhere solvable, if and only if the determinative equation is respectively densely solvable, normally solvable,* d-*normal, or everywhere solvable.*

This theorem shows that the properties of the determinative equation which were listed do not depend upon the choice of the closed direct complement L to the range $R(A)$.

The following theorem is useful in applications.

THEOREM 18.2. *Let equation* (A) *be* d-*normal. Then equation* (A,B) *is everywhere solvable if and only if* $R(A)^{\perp} \cap R(B)^{\perp} = \{\theta\}$.

PROOF. Recalling one of the first statements in this section, the condition of the theorem is necessary even for dense solvability. Conversely, if $R(A)^{\perp} \cap R(B)^{\perp} = \{\theta\}$, then equation (A,B) is densely solvable, and therefore, so is the determinative equation. Since $\dim L < \infty$, here the determinative equation is everywhere solvable, and, by Theorem 18.1, equation (A,B) is also everywhere solvable.

The theorem is proven.

THEOREM 18.3. *Assume that operator* A *has a dense domain in* E *and that the conditions of Theorem* 18.2 *are fulfilled. Then, for equation* (A,B), *everywhere solvability is stable to small perturbations of the operator* A.

PROOF. Since equation (A) is d-normal, the kernel $N(A^*) = R(A)^{\perp}$ is finite dimensional. By the condition of Theorem 18.2,

$N(A^*) \cap R(B)^{\perp} = \{\theta\}$. Now one can construct in F^* a closed direct complement M to $N(A^*)$, satisfying $M \supset R(B)^{\perp}$. By Theorem 16.2, equation (A + Q) is d-normal when $\| Q \|_{E \longrightarrow F}$ is small enough. To complete the proof of our theorem, it suffices to see that

$$N(A^* + Q^*) \cap R(B)^{\perp} = \{\theta\}.$$

As Observation 16.1 shows, for $\| Q^* \|_{F^* \longrightarrow E^*}$ (and hence for $\| Q \|_{E \longrightarrow F}$) small enough, the kernel $N(A^* + Q^*)$ intersects M only at zero. Then certainly $N(A^* + Q^*) \cap R(B)^{\perp} = \{\theta\}$, and the theorem is proven.

A similar proposition holds true for relativelly small perturbations of the operator A.

§ 19. INTEGRAL EQUATIONS

In this section and the one that follows, several types of important integral and differential equations, encountered in mathematical analysis, are considered. The discussion retains an illustrative character, which justifies why many basic facts are given without proofs.

1. FREDHOLM INTEGRAL EQUATIONS. Consider the Fredholm integral equation of second kind

$$x(t) - \int_0^1 Q(t,s)x(s)ds = y(t) \quad (0 \leq t \leq 1) \tag{19.1}$$

whose kernel $Q(t,s)$ is continuous in the square $0 \leq t,s \leq 1$. The operator

$$Tx(t) = \int_0^1 Q(t,s)x(s)ds$$

is linear and compact in the space $C[0,1]$. Indeed, the inequalities

$$|Tx(t)| \leq \max_{0\leq t,s\leq 1} |Q(t,s)| \int_0^1 |x(s)|ds$$

and

$$|Tx(t) - Tx(\tau)| \leq \max_{0\leq s\leq 1} |Q(t,s) - Q(\tau,s)| \int_0^1 |x(s)|ds$$

show that T takes any bounded subset of $C[0,1]$ into a uniformly bounded and equicontinuous family of functions, i.e., into a compact subset of $C[0,1]$. Moreover, it results from the same inequalities that T is also compact both as an operator from $L_1[0,1]$ into $C[0,1]$ and as an operator from $L_p[0,1]$ into $C[0,1]$, for $1 \leq p \leq \infty$. Therefore, (19.1) is a canonical Fredholm equation.

Since any set which is compact in $C[0,1]$ is also compact in $L_1[0,1]$, the operator T is compact as an operator in $L_1[0,1]$. The same argument shows that T is compact in all the spaces $L_p[0,1]$, $1 \leq p \leq \infty$.

Thus, *a Fredholm equation of the second kind is a canonical Fredholm equation in each of the spaces* $C[0,1]$, $L_p[0,1]$ $(1 \leq p \leq \infty)$.

Since a canonical Fredholm equation has index zero, we obtain the first basic result.

1°. *Equation* (19.1) *is everywhere solvable in the space* $L_1[0,1]$ *if and only if the homogeneous equation*

$$x(t) - \int_0^1 Q(t,s)x(s)ds = 0 \tag{19.2}$$

has only the trivial solution in $L_1[0,1]$.

This is a good place to make several useful remarks. If, given a function $y(t)$ from one of the spaces $C[0,1]$ or $L_p[0,1]$ $(1 \leq p \leq \infty)$, equation (19.1) is solvable in $L_1[0,1]$, then the solution automatically belongs to the same space as y, because it differs from $y(t)$ by the continuous function $Tx(t)$. In particular, the solutions of the homogeneous equation are always elements of $C[0,1]$.

If equation (19.1) is now everywhere solvable in one of the spaces E considered above, then it is solvable in the largest space $L_1[0,1]$, and hence, as the previous discussion shows, (19.1) is solvable in each of the spaces $C[0,1]$, $L_p[0,1]$ $(1 \leq p \leq \infty)$. To prove this statement, we observe that the space E is densely embedded into $L_1[0,1]$. Therefore, the everywhere solvability of (19.1) in E implies the dense solvability in $L_1[0,1]$. Now recall that a canonical Fredholm equation is normally solvable, i.e., equation (19.1) is everywhere solvable in $L_1[0,1]$.

Thus, one can make Property 1° even more precise.

1°a. *Equation* (19.1) *is everywhere solvable in one of the spaces* $C[0,1]$, $L_p[0,1]$ $(1 \leq p \leq \infty)$, *and hence in all of them, if and only if the homogeneous equation* (19.2) *has only the trivial solution in* $C[0,1]$ (*and so also in* $L_1[0,1]$).

It is easy to find the adjoint of the operator T in the space $L_1[0,1]$. The conjugate space of $L_1[0,1]$ is $L_\infty[0,1]$. Let $g(t) \in L_\infty[0,1]$. Then

$$\int_0^1 g(t)\int_0^1 Q(t,s)x(s)ds\,dt = \int_0^1 x(s)\int_0^1 Q(t,s)g(t)dt\,ds .$$

Therefore,

$$T^*g(s) = \int_0^1 Q(t,s)g(t)dt,$$

i.e., the adjoint operator is given by the transposed kernel.

Now let equation (19.1) be everywhere solvable in $L_1[0,1]$. Then the transposed equation

$$g(s) - \int_0^1 Q(t,s)g(t)dt = f(s)$$

is everywhere solvable in $L_\infty[0,1]$, and so in each of the spaces $C[0,1]$, $L_p[0,1]$ $(1 \leq p \leq \infty)$.

As a consequence, we get the remaining statements which, together with 1°, yield the classical Fredholm theorems.

2°. *Equation* (19.1) *is everywhere solvable in one of the spaces* $C[0,1]$, $L_p[0,1]$ $(1 \leq p \leq \infty)$, *if and only if the transposed equation*

$$g(s) - \int_0^1 Q(t,s)g(t)dt = f(s) \tag{19.3}$$

is everywhere solvable in one of these spaces.

3°. *The homogeneous equations* (19.2) *and*

$$g(s) - \int_0^1 Q(t,s)g(t)dt = 0 \tag{19.4}$$

have the same number of linearly independent continuous solutions.

4°. *Equation* (19.1) *is solvable for a given function* $y \in L_1[0,1]$ *if and only if the function* y *is orthogonal to all functions* g_i *of some basis for the space of solutions of equation* (19.4):

$$\int_0^1 y(s)g_i(s)ds = 0.$$

We make an additional remark concerning the Fredholm equation of the first kind

$$\int_0^1 Q(t,s)x(s)ds = y(t). \tag{19.5}$$

If its kernel is not degenerate (recall that a degenerate kernel is one of the form $Q(t,s) = \sum_1^N a_j(t)b_j(s)$), then the operator on the left-hand side is compact, but not of finite rank. As we mentioned on p. 7 , such an equation is not normally solvable, and so certainly not correctly solvable. Unfortunately, a lot of practical problems lead to the problem of solving such incorrectly solvable equations, and we have no chance in avoiding them. In recent years, there has been some progress in this problem, achieved by approaching equation (19.5) not in the entire space, but rather in some of its convex subsets, where one has a correct dependence upon the right-hand side.

2. VOLTERRA INTEGRAL EQUATIONS. A Volterra equation is an equation of the form

$$x(t) - \int_0^t Q(t,s)x(s)ds = y(t) \quad (0 \leq t \leq 1),$$

where $Q(t,s)$ is continuous in the triangle $0 \leq s \leq t \leq 1$. The operator

$$Tx(t) = \int_0^t Q(t,s)x(s)ds$$

acts again from $L_1[0,1]$ into $C[0,1]$. Indeed, if $t > \tau$, then

$$|Tx(t) - Tx(\tau)| \leq \underset{0\leq s\leq t\leq 1}{\text{Max}} |Q(t,s)| \int_\tau^t |x(s)|ds +$$

$$+ \underset{0\leq s\leq \tau\leq 1}{\text{Max}} |Q(t,s) - Q(\tau,s)| \, \|x\|_{L_1} \longrightarrow 0$$

as $|t - \tau| \longrightarrow 0$. However, T is not, in general, a compact operator from $L_1[0,1]$ into $C[0,1]$.

We have

$$|Tx(t)| \leq Q\int_0^t |x(s)|ds ,$$

where $Q = \underset{0\leq s\leq t\leq 1}{\text{Max}} |Q(t,s)|$. Then

$$|T^2x(t)| \leq Q\int_0^t |Tx(s_1)|ds_1 \leq Q^2\int_0^t\int_0^{s_1}|x(s)|ds\, ds_1 = Q^2\int_0^t (t-s)|x(s)|ds.$$

Successively applying such estimates, we obtain the inequality

$$|T^nx(t)| \leq Q^n\int_0^t [(t-s)^{n-1}/(n-1)!]|x(s)|ds \quad (n = 1,2,\ldots),$$

which shows that the operator

$$R = T + T^2 + \ldots + T^n + \ldots$$

acts from $L_1[0,1]$ into $C[0,1]$ and

$$\|Rx\|_{C[0,1]} \leq Q\int_0^t e^{Q(t-s)}|x(s)|ds \leq Qe^{Qt}\|x\|_{L_1} .$$

Consequently, the operator $I + R$ is bounded in $L_1[0,1]$ and is the inverse of the operator $I - T$. To conclude: *a Volterra equation is uniquely everywhere solvable in* $L_1[0,1]$, *and hence in all spaces* $C[0,1]$, $L_p[0,1]$ $(1 \leq p \leq \infty)$.

3. INTEGRAL EQUATIONS WHOSE KERNELS HAVE SUMMABLE SINGULARITIES. In equation (19.1), we drop the condition that $Q(t,s)$ be continuous, and assume that the kernel Q can be represented as

$$Q(t,s) = Q_1(t,s)k(t-s),$$

where the function $Q_1(t,s)$ is continuous in the square $0 \leq s,t \leq 1$, and the function $k(t)$ is even and summable on the segment $[-1,1]$. Then the operator

$$Tx(t) = \int_0^1 Q_1(t,s)k(t-s)x(s)ds \tag{19.6}$$

is bounded both in $C[0,1]$ and $L_1[0,1]$. Indeed,

$$|Tx(t)| \leq \max_{0\leq s,t\leq 1} |Q_1(t,s)| \, \|x\|_{C[0,1]} \int_0^1 |k(\sigma)| d\sigma \tag{19.7}$$

and

$$\int_0^1 |Tx(t)| dt \leq \max_{0\leq s,t\leq 1} |Q_1(t,s)| \int_0^1 \int_0^1 |k(t-s)| |x(s)| ds\, dt \leq$$
$$\leq \max_{0\leq s,t\leq 1} |Q_1(t,s)| \, \|x\|_{L_1[0,1]} \int_0^1 |k(\sigma)| d\sigma . \tag{19.8}$$

(To prove the last inequality, we used Fubini's theorem to justify changing the order of integration.)

Moreover, the operator T is compact in both spaces $C[0,1]$ and $L_1[0,1]$. Namely, inequalities (19.7) and (19.8) show that the norm of T in $C[0,1]$ and $L_1[0,1]$ is not larger than $Q_1 \|k\|_{L_1[-1,1]}$, where $Q_1 = \max_{0\leq t,s\leq 1} |Q(t,s)|$. Consequently, if we now approximate, in $L_1[-1,1]$, the function $k(t)$ by a sequence of even continuous functions $k_n(t)$, then the operators T_n, defined by (19.6) where k_n replaces k, converge to T in the operator norms of both $C[0,1]$ and $L_1[0,1]$. As we have seen in §19.1, T_n are compact, as operators with continuous kernels. Consequently, their norm limit T is also compact.

We deduce that

$$x(t) - \int_0^1 Q_1(t,s)k(t-s)x(s)ds = y(t) \tag{19.9}$$

is a canonical Fredholm equation in the spaces $C[0,1]$ and $L_1[0,1]$ (and, as the previous discussion shows, in all spaces $L_p[0,1]$).

As in §19.1, we conclude that if (19.9) is solvable in $L_1[0,1]$ for some given continuous right-hand side, the the solution is continuous. Further, the everywhere solvability of equation (19.9) in $C[0,1]$ implies its everywhere solvability in $L_1[0,1]$ and vice versa. This shows that the four statements that were formulated for equation (19.1) remain valid for equation (19.9).

The most frequently encountered examples of equations having the form (19.9) are integral equations whose kernels have a *weak singularity*. In this case $k(t) = |t|^{-\alpha}$, where $0 < \alpha < 1$.

We must emphasize that our discussion in §19.1 and 19.3 carries over without modification to integral equations given on any bounded closed subset of n-dimensional space. In such a situation, a kernel with a weak singularity is characterized by a factor $k(t) = |t|^{-\alpha}$, where t is an n-dimensional vector and $0 < \alpha < n$.

4. WIENER-HOPF EQUATIONS. [For a detailed analysis, see [51], [52].]

If an integral equation is considered on an infinite interval, then the integral operator may no longer be compact. As an example, take the convolution integral operator

$$Kx(t) = \int_{-\infty}^{+\infty} k(t-s)x(s)ds. \tag{19.10}$$

If we consider K in the space $L_1(-\infty,\infty) \equiv L_1$ and assume that the function $k \in L_1$, then it is a bounded operator. Indeed,

$$\int_{-\infty}^{\infty} |Kx(t)|dt \leq \int_{-\infty}^{\infty}\int_{-\infty}^{\infty} |k(t-s)||x(s)|ds\,dt =$$

$$= \int_{-\infty}^{\infty} |k(s)|ds \int_{-\infty}^{\infty} |x(s)|ds = \|k\|_{L_1}\|x\|_{L_1} .$$

However, K is not a compact operator (unless $k \equiv 0$). To see this, choose a sequence of functions $x_n(s) = \chi_{[0,a]}(s-n)$, where $\chi_{[0,a]}$ is the characteristic function of the segment $[0,a]$. Then the sequence $x_n(s)$ is bounded in L_1. Furthermore, $Kx_n(t) = \phi(t-n)$, where the function

$$\phi(t) = \int_{t-a}^{t} k(\sigma)d\sigma$$

is continuous and tends to zero at infinity. One can pick a in order to guarantee that $\phi(t) \not\equiv 0$. Then the sequence of shifted functions $\phi(t-n)$ cannot be compact in L_1. Assuming the contrary, the limit of a subsequence $\phi(t-n_k)$ $(n_k \longrightarrow \infty)$ must be zero, which contradicts the equality

$$\|\phi(t-n_k)\|_{L_1} = \|\phi(t)\|_{L_1} = \text{const.}$$

We remark that the sequence $\phi(t-n)$ is not compact in other spaces having a shift-invariant norm.

In other words, the equation

$$x(t) - \int_{-\infty}^{\infty} k(t-s)x(s)ds = y(t) \tag{19.11}$$

is not a canonical Fredholm equation in the space L_1. However, the problem of its solvability can be settled completely due to the special form of its kernel. This is done via the Fourier transform. Recall that the Fourier transform of a convolution equals the product of the Fourier transforms of its factors, and hence, by taking Fourier transforms, our equation becomes

$$X(\tau) - K(\tau)X(\tau) = Y(\tau),$$

where $X(\tau), K(\tau)$, and $Y(\tau)$ are the Fourier transforms of $x(t), k(t)$, and $y(t)$. Formally, we get

$$X(\tau) = (1 - K(\tau))^{-1} Y(\tau), \tag{19.12}$$

and this formula proves the uniqueness of the solution of (19.11).

In order that formula (19.12) yield a solution of our equation, the function $X(\tau)$ must be the Fourier transform of some L_1-function $x(t)$. Since such a transform is always a continuous function, a necessary condition is that

$$1 - K(\tau) \neq 0 \quad (-\infty < \tau < \infty). \tag{19.13}$$

It happens that this condition is also sufficient: by Wiener's theorem, (19.13) guarantees that the function $(1 - K(\tau))^{-1}$ can be written as $1 + K_1(\tau)$, where $K_1(\tau)$ is the Fourier transform of a function $k_1(t) \in L_1$. Now one can write the solution of equation (19.11) in the form

$$x(t) = y(t) + \int_{-\infty}^{\infty} k_1(t-s) y(s) ds \quad .$$

In other words, *equation* (19.11) *is uniquely everywhere solvable in the space* L_1 *if and only if condition* (19.13) *is fulfilled.*

The theory of the analogous equation on the half-line (which is called a *Wiener-Hopf equation*)

$$x(t) - \int_0^{\infty} k(t-s) x(s) ds = y(t) \quad (0 \leq t < \infty), \tag{19.14}$$

is considerably more complicated.

However, some important conclusions can be drawn on the basis of the analysis of the full-line equation. One can consider the space $L_1(-\infty, \infty)$ as the direct product $L_1(0,\infty) \times L_1(0,\infty)$: to each function $x(t) \in L_1(-\infty,\infty)$, we associate the pair of functions $x_1(t) = x(t)$, $x_2(t) = x(-t)$ $(0 \leq t < \infty)$ in $L_1(0,\infty)$. Obviously

$$\| x \|_{L_1(-\infty,\infty)} = \| x_1 \|_{L_1(0,\infty)} + \| x_2 \|_{L_1(0,\infty)} \cdot$$

Now one can write equation (19.11) as a system of two equations

$$\left. \begin{aligned} x_1(t) - \int_0^{\infty} k(t-s) x_1(s) ds - \int_0^{\infty} k(t+s) x_2(s) ds = y_1(t), \\ x_2(t) - \int_0^{\infty} k(-t-s) x_1(s) ds - \int_0^{\infty} k(-t+s) x_2(s) ds = y_2(t). \end{aligned} \right\} \tag{19.15}$$

Recalling the discussion above, if condition (19.13) is satisfied, then this system is uniquely everywhere solvable in the space $L_1(0,\infty) \times L_1(0,\infty)$.

Moreover, one can show that the operator

$$K_{12}x(t) = \int_0^\infty k(t+s)x(s)ds$$

is compact in $L_1(0,\infty)$. Indeed, one can verify (in the same way we did for operator (19.10)) that $\| K_{12}\| \leq \| k\|_{L_1(0,\infty)}$. The function $k(t)$ can be approximated arbitrarily well in $L_1(0,\infty)$ by a continuous function $h(t)$ which is equal to zero outside a large enough interval $[0,T]$. Then the operator $Hx(t) = \int_0^\infty h(t+s)x(s)ds$ takes any function $x \in L_1(0,\infty)$ into a continuous function which vanishes outside $[0,T]$. Moreover, H takes any bounded subset of $L_1(0,\infty)$ into a set of functions which is compact in $C[0,T]$, and so also in $L_1(0,\infty)$. Consequently, H is a compact operator, and so is K_{12}, being a norm limit of such operators.

The operator corresponding to system (19.15) may be written as a matrix

$$\begin{pmatrix} I - K_{11} & K_{12} \\ K_{21} & I - K_{22} \end{pmatrix} = \begin{pmatrix} I - K_{11} & 0 \\ 0 & I - K_{22} \end{pmatrix} + \begin{pmatrix} 0 & K_{12} \\ K_{21} & 0 \end{pmatrix} .$$

The operator on the left has, assuming (19.13), a bounded inverse, while the second operator on the right is compact. Now using Theorem 14.1, the first operator on the right is a Fredholm operator.

Therefore, *if condition* (19.13) *is satisfied, then both equation* (19.14) *and its transpose*

$$g(t) - \int_0^\infty k(-t+s)g(s)ds = f(t) \tag{19.16}$$

are Noetherian in the space $L_1(0,\infty)$.

In fact, one can show that *condition* (19.13) *is necessary and sufficient for equations* (19.14) *and* (19.16) *to be Noetherian.* If (19.13) is not fulfilled, then these equations are not even n-normal or d-normal.

For a more complete study of the Wiener-Hopf equation, more sophisticated analytic tools are needed.

The space $L_1(0,\infty)$ may be thought of as the subspace L_1^+ of L_1, consisting of all functions which are identically zero for $t < 0$. There is a natural projection P_+ onto this subspace: $P_+x(t) = x(t)$ for $t \geq 0$, $P_+x(t) = 0$ for $t < 0$. The integral operator K in

(19.14) acts from L_1^+ into L_1, and so equation (19.14) can be written in operator form:

$$Ax = x - P_+Kx = y \quad (x,y \in L_1^+).$$

One could try to solve this equation applying the method used above for solving the full-line equation. Firstly, we transform it to the form

$$x - Kx = y - P_-Kx \quad (P_- = I - P_+).$$

Secondly, we take the Fourier transform of this new equation and obtain

$$X(\tau) - K(\tau)X(\tau) = Y(\tau) + Z(\tau), \tag{19.17}$$

where $Z(\tau)$ is the Fourier transform of the function $-P_-Kx$ which contains the unknown function $x(t)$. Denote by Π_+ the projection operator which corresponds to P_+ under the Fourier transform. Now Π_+ takes the Fourier transform of a function x into the Fourier transform of its projection P_+x. One could try to eliminate the function $Z(\tau)$ from equation (19.17) by simply applying Π_+ to both sides. However, in this way we would destroy the multiplicative structure of the left-hand side and thus kill the possibility of finding $X(\tau)$ immediately. But here we are aided by a remarkable state of affairs. Namely, it is true that, under certain conditions, the function $(1 - K(\tau))^{-1}$ can be factored as

$$(1 - K(\tau))^{-1} = G_+(\tau)G_-(\tau) , \tag{19.18}$$

where both factors are Fourier transforms of functions, from L_1^+ and L_1^- ($= P_-L_1$) respectively, and $G_+^{-1}(\tau)$ is also the Fourier transform of an L_1^+-function. The one can rewrite equation (19.17) as

$$G_+^{-1}(\tau)X(\tau) = G_-(\tau)Y(\tau) + G_-(\tau)Z(\tau) .$$

Now the left-hand side is a product of Fourier transforms of functions from L_1^+, and hence is itself the Fourier transform of a function from L_1^+ (a convolution of functions from L_1^+ is again a function from L_1^+). Consequently, when we apply the operator Π_+, the left-hand side does not change. Moreover, the function $G_-(\tau)Z(\tau)$ is the Fourier transform of a function from L_1^-, and so $\Pi_+G_-(\tau)Z(\tau) = 0$. We get

$$G_+^{-1}(\tau)X(\tau) = \Pi_+G_-(\tau)Y(\tau)$$

and, finally

$$X(\tau) = G_+(\tau)\Pi_+G_-(\tau)Y(\tau) . \tag{19.19}$$

The existence of a factorization (19.18) is determined by the number

$$\nu = -\frac{1}{2\pi}\int_{-\infty}^{\infty} d_\tau \arg(1 - K(\tau)) . \tag{19.20}$$

In this way, the following results were obtained.

1°. *Equation* (19.14) *is uniquely everywhere solvable in* L_1 *if and only if condition* (19.13) *is fulfilled and* $\nu = 0$.

2°. *If condition* (19.13) *is fulfilled and* $\nu > 0$, *then equation* (19.14) *is everywhere solvable in* L_1 *and* n-*normal. In this case, the dimension* $n(A)$ *of the kernel equals* ν, *the defect* $d(A)$ *equals* 0, *and so the index* $\operatorname{ind}(A) = \nu$.

3°. *If condition* (19.13) *is fulfilled and* $\nu < 0$, *then equation* (19.14) *is Noetherian. Also,* $n(A) = 0$, $d(A) = -\nu$, *and the index is again* $\operatorname{ind}(A) = \nu$.

In the last case, equation (19.14) is solvable if the right-hand side is orthogonal to all solutions of the transposed equation

$$g(s) - \int_0^\infty k(t-s)g(t)dt = 0 , \tag{19.21}$$

i.e., if

$$\int_0^\infty y(s)g_i(s)ds = 0 \quad (i = 1,\dots,-\nu) ,$$

where $\{g_i(s)\}$ is a basis in the space of solutions of (19.21).

The explicit formula (19.19) for the solution of equation (19.14) allows one to prove all the claims listed above for the spaces $L_p(0,\infty)$ $(1 \leq p \leq \infty)$, the space of continuous and bounded functions on $[0,\infty)$, and the spaces of uniformly continuous functions, continuous functions having a limit at ∞, and continuous functions which tend to 0 at ∞.

In [51] a theory similar to that discussed above, but more complicated, is developed for systems of integral equations on the half-line and having kernels depending on the difference of arguments.

Recently, these methods were extended to more general classes of equations, with so-called pseudo-differential operators (see [50]).

5. SINGULAR INTEGRAL EQUATIONS. [For a detailed analysis, see the book [53].] Integral operators loose the property of being compact when their kernels have singularities which are not summable. Here one should remark, at the onset, that even the definition of the integral operator must be generalized and made more rigorous when under the integral sign there appear functions which are not summable. We shall analyze the problems which must be faced in this situation for the example of a Cauchy singular integral operator.

Let Γ be a closed curve in the complex plane. Assume that Γ does not intersect itself, and is given by an equation $t = t(\sigma)$ $(0 \leqq \sigma \leqq 1)$, where $t(\sigma)$ is a continuously differentiable function whose derivative is everywhere different from zero. The Cauchy operator acts on complex-valued functions $x(t)$ defined on Γ by the formula

$$\int_\Gamma \frac{x(\tau)d\tau}{\tau - t} .$$

If the point $t = z \notin \Gamma$, then the integral exists, and it defines two functions $\Phi^+(z)$ and $\Phi^-(z)$, holomorphic inside and outside contour Γ respectively. If the point $t \in \Gamma$, then the integral no longer makes sense as a Lebesgue integral. However, one can try to consider its principal value

$$\text{v.p.} \int_\Gamma \frac{x(\tau)d\tau}{\tau - t} = \lim_{\varepsilon \to 0} \int_{\Gamma_\varepsilon} \frac{x(\tau)}{\tau - t} d\tau \tag{19.22}$$

where Γ_ε is obtained from the contour Γ by deleting its intersection with a neighborhood $|z - t| < \varepsilon$ of t. One can check easily that

$$\text{v.p.} \int_\Gamma \frac{d\tau}{\tau - t} = \pi i ,$$

and so our operator is already defined on constant functions. Moreover, if the function $x(t)$ satisfies a Hölder condition $(|x(t) - x(\tau)| \leqq \leqq C|t - \tau|^\alpha,\ 0 < \alpha \leqq 1)$, then

$$\text{v.p.} \int_\Gamma \frac{x(\tau)d\tau}{\tau - t} = \text{v.p.} \int_\Gamma \frac{x(\tau) - x(t)}{\tau - t} d\tau + \text{v.p.} \int_\Gamma \frac{x(t)}{\tau - t} d\tau =$$

$$= \int_\Gamma \frac{x(\tau) - x(t)}{\tau - t} d\tau + \pi i x(t) .$$

Indeed, the function under the first integral sign has a weak singularity (of order $|\tau - t|^{\alpha-1}$) and hence the corresponding integral exists in the usual sense. Therefore, operator (19.22) is defined on all Hölder spaces $C_\alpha(\Gamma)$ $(0 < \alpha \leqq 1)$.

For the following discussion, it is convenient to introduce the operator

$$Sx(t) = \frac{1}{\pi i} \text{v.p.} \int_\Gamma \frac{x(\tau)}{\tau - t} d\tau .$$

The important Plemelj-Privalov theorem states that *the operator* S *is bounded in all spaces* $C_\alpha(\Gamma)$ $(0 < \alpha < 1)$.

The operator S has the remarkable property that

$$S^2 = I . \tag{19.23}$$

The proof of this identity rests upon the Sohockii-Plemelj formulas for the limit values of Cauchy-type integrals: if the function $x(t)$ satsfies a Hölder condition, then for the functions $\Phi^+(z)$ and $\Phi^-(z)$ introduced above the limit values on the contour Γ, $\Phi^+(t)$ and $\Phi^-(t)$, exist and are given by

$$\Phi^+(t) = \pi i x(t) + \text{v.p.} \int_\Gamma \frac{x(\tau)}{\tau - t} d\tau ,$$

$$\Phi^-(t) = - \pi i x(t) + \text{v.p.} \int_\Gamma \frac{x(\tau)}{\tau - t} d\tau .$$

Therefore, $x(t) = \frac{1}{2\pi i}(\Phi^+(t) - \Phi^-(t))$. Furthermore, as one may check easily, $S\Phi^+(t) = \Phi^+(t)$, $S\Phi^-(t) = -\Phi^-(t)$, and now identity (19.23) is plain.

Now consider the simplest singular integral equation

$$Ax(t) \equiv a(t)x(t) + \frac{b(t)}{\pi i} \text{v.p.} \int_\Gamma \frac{x(\tau)}{\tau - t} d\tau = y(t) . \tag{19.24}$$

If a function $\phi(t)$ satisfies a Holder condition with exponent μ, then the operator of multiplication by $\phi(t)$ is bounded in all Holder spaces $C_\alpha(\Gamma)$ with $\alpha \leq \mu$. Using all these facts, we can conclude that given $a,b \in C_\mu(\Gamma)$ $(0 < \mu \leq 1)$, the operator A is bounded in all spaces $C_\alpha(\Gamma)$ with $\alpha \leq \mu$.

The second term in the left-hand side of our equation has the form bS, where b is the operator of multiplication by $b(t)$. Such multiplication operators and the operator S satisfy the following important commutation formula:

$$bSx - S(bx) = \frac{1}{\pi i} \int_\Gamma \frac{b(t) - b(\tau)}{\tau - t} x(\tau) d\tau .$$

Since $b \in C_\mu(\Gamma)$, the kernel $\frac{b(t) - b(\tau)}{\tau - t}$ has a weak singularity. In fact, it turns out that it defines a compact integral operator in all spaces $C_\alpha(\Gamma)$ with $\alpha < \mu/2$. For what follows, we remark that the same property is enjoyed by the more general operator

$$\int_\Gamma \frac{Q(t,\tau) - Q(t,t)}{\tau - t} x(\tau) d\tau , \tag{19.25}$$

if the function $Q(t,\tau)$ is of Hölder class in both variables.

Therefore, the operator $T_b = bS - Sb$ is compact in $C_\alpha(\Gamma)$, $\alpha < \mu/2$. This allows us to construct explicitely a right and a left regularizer for equation (19.24).

Assume that $a,b \in C_\mu(\Gamma)$ and

$$a^2(t) - b^2(t) \neq 0 \quad (t \in \Gamma).$$

Then

$$(a - bS)(a + bS) = a^2 + abS - bSa - bSbS = a^2 + bT_a - b^2 + bT_bS =$$
$$= a^2 - b^2 + T ,$$

where $T = bT_a + bT_bS$ is a compact operator in $C_\alpha(\Gamma)$ with $\alpha < \mu/2$. Consequently, the operator

$$B = \frac{1}{a^2 - b^2} (a - bS)$$

is both a right and left regularizer for equation (19.24).

Our main conclusion is: *for functions*

$$a(t),b(t) \in C_\mu(\Gamma) \quad (0 < \mu \leqq 1),$$

equation (19.24) *is Noetherian in all spaces* $C_\alpha(\Gamma)$ *with* $\alpha < \mu/2$.

A remarkable fact is that one can compute explicitely the index of equation (19.24):

$$\text{ind}(A) = \frac{1}{2\pi} \int_\Gamma d_t \arg \frac{a(t) + b(t)}{a(t) - b(t)}$$

in all spaces $C_\alpha(\Gamma)$ with $\alpha < \mu/2$. [The properties of singular integral equations which have been discussed above were discovered by F. Noether.]

Using general theorems concerning the preservation of the Noetherian character and the invariance of the index under compact perturbations, one can immediately consider more general equations than (19.24), namely equations of the form

$$a(t)x(t) + \frac{1}{\pi i} \text{ v.p.} \int_\Gamma \frac{Q(t,\tau)}{\tau - t} x(\tau)d\tau = y(t) . \tag{19.26}$$

Transforming this equation to the form

$$a(t)x(t) + \frac{Q(t,t)}{\pi i} \text{ v.p.} \int_\Gamma \frac{x(\tau)}{\tau - t} d\tau + \frac{1}{\pi i} \int_\Gamma \frac{Q(t,\tau) - Q(t,t)}{\tau - t} x(\tau)d\tau =$$
$$= y(t),$$

we see that it contains a term which does not appear in (19.24) and involves the operator (19.25) which is compact in $C_\alpha(\Gamma)$ with $\alpha < \mu/2$. Then the following statement holds true.

Let $a \in C_\mu(\Gamma)$, *let the kernel* $Q(t,\tau)$ *satisfy Hölder's condition with exponent* μ *in the two variables, and let*

$$a^2(t) - Q^2(t,t) \neq 0 \qquad (t \in \Gamma).$$

Then equation (19.26) *is Noetherian in all spaces* $C_\alpha(\Gamma)$ *with* $\alpha < \mu/2$, *and its index is given by the formula*

$$\nu = \frac{1}{2\pi}\int_\Gamma d_t \arg \frac{a(t) + Q(t,t)}{a(t) - Q(t,t)} .$$

We remark that there exists a rich theory of singular integral equations with Cauchy kernel on an open contour Γ, which we do not touch here.

As an example of a real singular integral equation, consider the equation

$$a(s)x(s) + \frac{1}{2\pi}\,\text{v.p.}\int_0^{2\pi} Q(s,\sigma)\operatorname{ctg}\frac{\sigma - s}{2}\, x(\sigma)d\sigma = y(s) \qquad (19.27)$$

with Hilbert kernel, where $a(s)$ and $Q(s,\sigma)$ are real, 2π-periodic functions of s and σ, which satisfy in these variables Hölder's condition with exponent μ.

Let us perform the changes

$$t = e^{is}, \quad \tau = e^{i\sigma}, \quad \widetilde{Q}(t,\tau) = Q(-i \ln t, -i \ln \tau),$$

$$\widetilde{a}(t) = a(-i \ln t), \quad \widetilde{y}(t) = y(-i \ln t), \quad \widetilde{x}(t) = x(-i \ln t).$$

Then $\operatorname{ctg}\frac{\sigma - s}{2} = i(\frac{2\tau}{\tau - t} - 1)$ and $d\sigma = \frac{1}{i}\frac{d\tau}{\tau}$, and we get the equation

$$\widetilde{a}(t)\widetilde{x}(t) + \frac{1}{\pi}\,\text{v.p.}\int_\Gamma \widetilde{Q}(t,\tau)\widetilde{x}(\tau)\frac{d\tau}{\tau - t} - \frac{1}{2\pi}\int_\Gamma \widetilde{Q}(t,\tau)\widetilde{x}(\tau)\frac{d\tau}{\tau} = \widetilde{y}(t) , \qquad (19.28)$$

where Γ is the unit circle in the complex plane. This equation differs from (19.26) by the presence of the supplementary operator

$$-\frac{1}{2\pi}\int \widetilde{Q}(t,\tau)\widetilde{x}(\tau)\frac{d\tau}{\tau} ,$$

which is compact in all spaces $C_\alpha(\Gamma)$ with $\alpha \leq \mu$. Again, the stability of the Noetherian character and the invariance of the index under compact perturbations show that *equation* (19.28), *and so equation* (19.27) *are Noetherian in all spaces* $C_\alpha(\Gamma)$ $(C_\alpha^p[0,2\pi])$ *with* $\alpha < \mu/2$, *and that their index is given by the formula*

$$\nu = \frac{1}{2\pi}\int_0^{2\pi} d_s \arg \frac{a(s) + iQ(s,s)}{a(s) - iQ(s,s)} .$$

(Here $C_\alpha^p[0,2\pi]$ denotes the space of all periodic functions on $[0,2\ ]$ which satisfy the Hölder condition with exponent α).

We have examined the solvability of singular integral equations in Hölder spaces. A similar theory exists for the spaces $L_p(\Gamma)$ $(1 < p < \infty)$.

§ 20. DIFFERENTIAL EQUATIONS

1. BOUNDARY VALUE PROBLEMS FOR ORDINARY DIFFERENTIAL EQUATIONS. Consider the differential equation

$$\ell x(t) = x^{(r)}(t) + p_1(t)x^{(r-1)}(t) + \ldots + p_r(t)x(t) = y(t), \quad (0 \leqq t \leqq 1) \tag{20.1}$$

with continuous coefficients $p_k(t)$ on $[0,1]$. The associated homogeneous equation

$$x^{(r)}(t) + p_1(t)x^{(r-1)}(t) + \ldots + p_r(t)x(t) = 0$$

has r linearly independent, r-times continuously differentiable solutions $x_1(t),\ldots,x_r(t)$. The corresponding Wronskian determinant

$$\Delta(t) = \begin{vmatrix} x_1(t) & \ldots & x_r(t) \\ x_1'(t) & \ldots & x_r'(t) \\ \ldots & \ldots & \ldots \\ x_1^{(r-1)}(t) & \ldots & x_r^{(r-1)}(t) \end{vmatrix} \neq 0 \quad (0 \leqq t \leqq 1).$$

Denote by $\Delta_k(t)$ the cofactor of the k-th element of the last row of the Wronskian determinant. Then the function

$$x(t) = \sum_{k=1}^{r} \int_0^t \frac{x_k(t)\Delta_k(\tau)}{\Delta(\tau)} y(\tau)d\tau \tag{20.2}$$

is a solution of equation (20.1). Moreover,

$$x^{(i)}(t) = \sum_{k=1}^{r} \int_0^t \frac{x_k^{(i)}(t)\Delta_k(\tau)}{\Delta(\tau)} y(\tau)d\tau \quad (i = 0,1,\ldots,r-1)$$

and, in particular

$$x(0) = x^{(1)}(0) = \ldots = x^{(r-1)}(0) \ . \tag{20.3}$$

Now let E be either the space $C[0,1]$ or $L_p[0,1]$ $(1 \leq p \leq \infty)$, and consider in E the operator $A_0x = x$, defined on all functions having $r-1$ absolutely continuous derivatives and a derivative of order

r, all of which belong to E and satisfy condition (20.3). Then *the equation* $A_0x = y$ *is uniquely everywhere solvable in the space* E. (The uniqueness is a corollary of the uniqueness of the solution of the Cauchy problem for ordinarry differential equations.)

One could consider a more general operator A_C, which is defined on all functions having $r-1$ absolutely continuous derivatives and a derivative of order r, all of which belong to E, and acts from E into the space $E \times \mathbb{R}^r = F$ according to the formula

$$A_C x = \{\ell x(t);\ x(0),\ldots,x^{(r-1)}(0)\} \ .$$

In this case the following holds.

Equation $A_C x = y$, *corresponding to the Cauchy problem for equation* (20.1), *is uniquely everywhere solvable in the space* F.

Indeed, the solution can be written explicitely as

$$x(t) = \sum_{k=1}^{r} \int_0^t \frac{x_k(t)\Delta_k(\tau)}{\Delta(\tau)}\, y(\tau)d\tau + \sum_{k=1}^{r} c_k^0\, x_k(t), \tag{20.4}$$

where c_k^0 satisfy the linear system

$$\sum_{k=1}^{r} c_k^0 x_k^{(i)}(0) = x_0^{(i)} \qquad (i = 0,1,\ldots,r-1)$$

with determinant $\Delta(0) \neq 0$. Here $\{x_0^{(i)}\}$ is an arbitrarily given vector from $\mathbb{R}^r$.

Since the inverse of the operator A_C is bounded, we see that A_C is closed as an operator from E to F. The general theory shows that one can introduce the graph norm on the domain of the operator A_C and then this domain becomes a Banach space. In our case, the graph norm is equivalent, for example, to the norm

$$\|x\|_{W_E^r(0,1)} = \sum_{k=0}^{r-1} \sup_{0\le t\le 1} |x^{(k)}(t)| + \|x^{(r)}\|_E \ .$$

Obviously, if $E = C[0,1]$, then $W_E^r(0,1) = C^r[0,1]$. If $E = L_p[0,1]$, then $W_E^r(0,1)$ is usually denoted by $W_p^r(0,1)$.

The operator A_C is a one-to-one and bicontinuous map from the space $W_E^r(0,1)$ onto the space $F = E \times \mathbb{R}^r$.

Now let us try to solve the following problem: find a solution $x(t)$ of equation (20.1) that satisfies the additional conditions

$$U_i(x) = \phi_i, \quad i = 1,2, \ldots ,m, \tag{20.5}$$

where $U_i(x)$ is a given system of m linear independent functionals on the space $W_E^r(0,1)$. To treat the problem (20.1), (20.5) using operator theory, consider the operator A_U with domain $W_E^r(0,1)$ and given by the formula

$$A_U(x) = (\ell x(t); U(x)), \quad U(x) = \{U_1(x), \dots, U_m(x)\} .$$

Therefore, A_U acts from E into $F_1 = E \times \mathbb{R}^m$.

The space $W_E^r(0,1)$ can be decomposed into the direct sum $\mathcal{D}(A_0) \oplus N(\ell)$, where $N(\ell)$ is the finite dimensional space of all solutions of the homogeneous equation $\ell x = 0$. The restriction A_U^0 of A_U to $\mathcal{D}(A_0)$ maps $\mathcal{D}(A_0)$ one-to-one onto the set G of elements in F_1 which have the form $(y, UA_0^{-1}(y))$ $(y \in E)$. Since the functionals U_i are continuous, G is a subspace of F_1. Moreover, G does not contain elements of the form (θ, ϕ), and so the space F_1 decomposes into the direct sum $F_1 = G \oplus (\{\theta\} \times \mathbb{R}^m)$. The operator A_U takes $N(\ell)$ into the subspace $\{\theta\} \times \mathbb{R}^m$. We see that the kernel of A_U coincides with the kernel of its restriction to $N(\ell)$. The range of A_U is the direct sum of G and a finite dimensional space, and thus is closed. The defect of the operator A_U equals the defect of its restriction to $N(\ell)$, when the latter is considered as an operator from this r-dimensional space into the m-dimensional space $\{\theta\} \times \mathbb{R}^m$. As we have noted in the Introduction, the index of an operator acting from an r-dimensional space into an m-dimensional space is $r-m$.

Our conclusion is: *the equation* $A_U x = (y, \phi)$, *corresponding to the problem* (20.1), (20.5) *is Noetherian, and its index equals* $r-m$.

As a corollary, we get a necessary condition for the unique and everywhere solvability of problem (20.1), (20.5), namely, the number m of additional conditions should equal r.

When this condition does hold, a sufficient condition for the unique, and so everywhere solvability of our problem is that the system of equations

$$c_1 U_1(x_1) + \dots + c_r U_1(x_r) = 0,$$
$$\dots\dots\dots\dots\dots\dots\dots\dots$$
$$c_1 U_r(x_1) + \dots + c_r U_r(x_r) = 0,$$

have only the trivial solution.

Therefore, *problem* (20.1), (20.5) *is uniquely and everywhere solvable if and only if* $m = r$ *and the determinant* $|U_i(x_j)|$ *is different from zero.*

In order to describe the right-hand sides for which problem (20.1), (20.5) is solvable, it is natural to pass to the adjoint operator. If we consider the operator (20.2) inverse to the operator A_0, we see that by transposing its kernel, the factors $\Delta_k(t)/\Delta(t)$ emerge outside the integral sign, and the only thing we can guarantee about these factors is their continuity. The domain of the adjoint operator A_0^* will have a rather strange description, and the operator A_0^* will not be defined by a differential expression of the same classical type as the expression ℓx. This difficulty is eliminated when one assumes additionally that the coefficients are smooth, or, more precisely, that $p_k(t)$ has $r-k$ continuous derivatives. Then one can integrate by parts to get the identity

$$\int_0^1 \ell x(t)g(t)dt = P(\xi(1),\eta(1)) - P(\xi(0),\eta(0)) + \int_0^1 x(t)\ell^{\dagger}g(t)dt, \qquad (20.6)$$

where the expression

$$\ell^{\dagger}g(t) \equiv (-1)^r g^{(r)}(t) + (-1)^{r-1}(p_1(t)g(t))^{(r-1)} + \dots + p_r(t)g(t)$$

becomes, upon opening the brackets, an ordinary, linear differential expression with continuous coefficients. The latter is called the differential expression *adjoint* (or *transpose*) to ℓ. The expression $P(\xi(i),\eta(i))$ is a bilinear form of the r-dimensional vectors

$$\begin{aligned} \xi(i) &= (x(i),x^{(1)}(i),\dots,x^{(r-1)}(i)), \\ \eta(i) &= (g(i),g^{(1)}(i),\dots,g^{(r-1)}(i)), \end{aligned} \qquad (i = 0,1).$$

Its explicit form shows that $P(\xi,\eta)$ is nondegenerate, i.e., $P(\xi,\eta) = 0$ for all η and fixed ξ implies $\xi = 0$. Based on this observation, we can establish the following important auxilliary proposition.

Let K *denote the set of values taken by the operator* A_0 *acting in* $L_1[0,1]$ *on the functions of* $\mathcal{D}(A_0)$ *which vanish together with their derivatives of order* $1,\dots,r-1$ *at the extremities of the segment* $[0,1]$. *Then the orthogonal complement of* K *in* $L_\infty[0,1]$ *is the space* $N(\ell^{\dagger})$ *of all the solutions of the adjoint homogeneous equation* $\ell^{\dagger}g = 0$.

PROOF. Let the function $y \in L_1[0,1]$ be orthogonal to $N(\ell^{\dagger})$, and let $x = A_0^{-1}y$. Then, by (20.6),

$$0 = \int_0^1 ygdt = \int_0^1 (\ell x)gdt = P(\xi(1),\eta(1))$$

for all $g \in N(\ell^{\dagger})$. As the function g runs over all $N(\ell^{\dagger})$, the vector $\eta(1) = (g(1),g^{(1)}(1),\dots,g^{(r-1)}(1))$ runs over the entire space $\mathbf{R}^r$. Since the form $P(\xi,\eta)$ is nondegenerate, the previous equality implies that $\xi(1) = 0$. Consequently,

$$y = \ell x \quad \text{and} \quad x(0) = x^{(1)}(0) = \ldots = x^{(r-1)}(0) = 0$$

(by construction), and

$$x(1) = x^{(1)}(1) = \ldots = x^{(r-1)}(1) = 0$$

(as we proved), i.e., the orthogonal complement of $N(\ell^{\dagger})$ in $L_1[0,1]$ is the set K of elements of the form ℓx, where the function x satisfies the boundary conditions given above. Since $\dim N(\ell^{\dagger}) < \infty$, $N(\ell^{\dagger})$ itself is the orthogonal complement of K in $L_\infty[0,1]$, Q.E.D.

Now turning to the description of the operator adjoint to the operator A_U with additional conditions (20.5), we restrict ourselves only to the case of classical boundary conditions

$$B_j(x) = \phi_j \;, \quad j = 1,2,\ldots,m, \tag{20.7}$$

where $\{B_j(x)\}$ is a system of linearly independent, linear forms of the $2r$ variables

$$x(0), \ldots, x^{(r-1)}(0), x(1), \ldots, x^{(r-1)}(1) \;.$$

We denote the operator A_U by A_B, and consider it in the largest space $L_1[0,1]$. A_B takes values in the space $L_1[0,1] \times \mathbb{R}^m$, whose dual is the space $L_\infty[0,1] \times \mathbb{R}^m$. The adjoint operator acts from $L_\infty[0,1] \times \mathbb{R}^m$ into $L_\infty[0,1]$.

Let the pair (g,ψ) $(g \in L_\infty[0,1], \psi \in \mathbb{R}^m)$ be an element of the domain of the conjugate operator. Then for $x \in \mathcal{D}(A_B)$,

$$\int_0^1 (\ell x(t))g(t)dt + \textstyle\sum_{j=1}^m B_j(x)\psi_j = \displaystyle\int_0^1 x(t)f(t)dt, \tag{20.8}$$

where $f \in L_\infty[0,1]$. If $x(t)$ and all its derivatives of orders less than or equal to $r-1$ are equal to zero at the extremities of the segment $[0,1]$, then (20.8) becomes simpler

$$\int_0^1 (\ell x(t))g(t)dt = \int_0^1 x(t)f(t)dt \;. \tag{20.9}$$

Denote by $A_0^{\dagger}$ the operator constructed from the differential expression $\ell^{\dagger}$ in the same way that A_0 was constructed from ℓ. Then any $f \in L_\infty[0,1]$ can be represented as $f = A_0^{\dagger}h$, with $h \in W_\infty^r(0,1)$. If we substitute this expression for f in (20.9) and integrate by parts, using the boundary conditions satisfied by $x(t)$, we get

$$\int_0^1 (\ell x(t))g(t)dt = \int_0^1 (\ell x(t))h(t)dt.$$

Therefore, the function $g(t)$ differs from $h(t) \in W_\infty^r(0,1)$ by a function which is orthogonal to all $\ell x(t)$, and thus, as we have seen above, $g(t) - h(t)$ is a solution of the homogeneous, adjoint differential equation. Since any such solution is r times continuously differentiable, we have $g \in W_\infty^r(0,1)$ and $f = \ell^\dagger h = \ell^\dagger g$. Now apply the integration by parts formula (20.6) to an arbitrary function $x \in W_1^r(0,1)$ and to g. However, we shall transform the terms which are not under the integral sign to a special form. To this purpose, we consider, along with the m linearly independent forms B_i $(i = 1,\dots,m)$ of the $2r$ variables

$$x(0),\ x^{(1)}(0),\ \dots,\ x^{(r-1)}(0),\ x(1),\ x^{(1)}(1),\ \dots,\ x^{(r-1)}(1),$$

$2r-m$ additional forms $S_1,\dots,S_{2r-m}$, such that the forms $B_1,\dots,B_m$, $S_1,\dots,S_{2m-r}$ are linearly independent. In other words, we complete the $2r$-dimensional vectors given by the coefficients of $B_1,\dots,B_m$, to a basis in the $2r$-dimensional space. Now take new independent variables to be the values of the forms $-B_1,\dots,-B_m,S_1,\dots,S_{2r-m}$, express $x(0),\dots,x^{(r-1)}(0),x(1),\dots,x^{(r-1)}(1)$ linearly in terms of these new variables, and then substitute the resulting expressions in (20.6). Grouping the like terms relative to the new variables, we can rewrite (20.6) in the form

$$\int_0^1 (\ell x(t))g(t)dt + \sum_{j=1}^m B_j(x)S_j^\dagger(g) =$$

$$= \sum_{k=1}^{2r-m} S_k(x)B_k^\dagger(g) + \int_0^1 x(t)\ell^\dagger g(t)dt, \qquad (20.10)$$

where $S_1^\dagger(g),\dots,S_m^\dagger(g),B_1^\dagger(g),\dots,B_{2r-m}^\dagger(g)$ are linear forms in the variables $g(0),\dots,g^{(r-1)}(0),g(1),\dots,g^{(r-1)}(1)$. Since the form $P(\xi,\eta)$ is nondegenerate, these linear forms are clearly linearly independent.

Now compare equalities (20.8) and (20.10). We know already that $f(t) = \ell^\dagger g(t)$. As x runs over $\mathcal{D}(A_B) = W_1^r(0,1)$, the vector $\{B_1(x),\dots,B_m(x),S_1(x),\dots,S_{2r-m}(x)\}$ runs over the entire $2r$-dimensional space $\mathbb{R}^{2r}$. Consequently,

$$B_1^\dagger(g) = \dots = B_{2r-m}^\dagger(g) = 0, \quad S_1^\dagger(g) = \psi_1,\ \dots,\ S_m^\dagger(g) = \psi_m\ . \qquad (20.11)$$

Conversely, if a function $g \in W_\infty^r(0,1)$ satisfies (20.11), then (20.8) holds, i.e., $g \in \mathcal{D}(A_B^*)$.

To summarize: *the domain of the operator adjoint to the operator*

A_B *corresponding to the boundary value problem* (20.1), (20.7) *consists of all elements of the form* $(g(t), S_1^\dagger(g), \ldots, S_m^\dagger(g))$, *where* g *is an arbitrary function from* $W_\infty^r(0,1)$ *satisfying the boundary conditions*

$$B_1^\dagger(g) = \ldots = B_{2r-m}^\dagger(g) = 0.$$

The operator A_B^* *itself is given by the formula*

$$A_B^*(g(t), S_1^\dagger(g), \ldots, S_m^\dagger(g)) = \ell^\dagger g.$$

Our concluding result is a consequence of general theorems concerning normal solvability.

Given a function $y(t) \in L_1[0,1]$ *and a vector* $\phi \in \mathbb{R}^m$, *the boundary value problem* (20.1), (20.7) *has a solution if and only if the orthogonality condition*

$$\int_0^1 y(t)g(t)dt + \sum_{i=1}^m \phi_i S_i^\dagger(g) = 0$$

is fulfilled for all solutions g *of the homogeneous, adjoint differential equation* $\ell^\dagger g = 0$.

Here, strictly speaking, we conclude our discussion of the solvability of boundary value problems for ordinary differential operators with continuous coefficients on a finite interval. For equations on unbounded intervals or equations with singular coefficients, difficult problems arise in connection with the asymptotic behaviour of the solutions at infinity or in the neighborhood of singular points. These problems are not discussed here.

In conclusion, we remark that all the problems that we have considered have real coefficients. The extension of our analysis to the case of complex coefficients presents no problems.

2. BOUNDARY VALUE PROBLEMS FOR ELLIPTIC DIFFERENTIAL EQUATIONS. Consider first the differential expression of order r with constant coefficients

$$A(D) = \sum_{|\alpha| \le r} a_\alpha(t) D^\alpha ,$$

where $\alpha = (\alpha_1, \ldots, \alpha_n)$ is an integral multi-index, $|\alpha| = \alpha_1 + \ldots + \alpha_n$,

$$D^\alpha = D_1^{\alpha_1} D_2^{\alpha_2} \ldots D_n^{\alpha_n} , \quad D_k = i \frac{\partial}{\partial t_k} \quad (k = 1,2,\ldots,n) ,$$

$t = (t_1, \ldots, t_n)$ is a point of the n-dimensional space $\mathbb{R}^n$, and a_α are given complex numbers.

By definition, the *principal part* of the differential expression $A(D)$ is the expression

$$A'(D) = \sum_{|\alpha|=r} a_\alpha D^\alpha .$$

To each differential expression, one can associate the homogeneous polynomial in n variables $(\zeta_1,\dots,\zeta_n) = \zeta$:

$$A'(\zeta) = \sum_{|\alpha|=r} a_\alpha \zeta^\alpha ,$$

where $\zeta^\alpha = \zeta_1^{\alpha_1}\zeta_2^{\alpha_2} \cdots \zeta_n^{\alpha_n}$.

A differential expression $A(D)$ is called *elliptic* if $A'(\zeta) \neq 0$ for all $\zeta \neq 0$.

If $n > 2$, then the order of an elliptic expression is an even number: $r = 2m$. This is not necessarily so when $n = 2$.

For fixed, linearly independent vectors, ζ' and ζ'', the function $A'(\zeta' + \tau\zeta'')$ does not vanish for τ real. As a polynomial in τ, this function has r complex roots. The differential expression $A(D)$ is said to be *properly elliptic* (or *to satisfy the root condition*) if $r = 2m$ and for fixed, linearly independent ζ' and ζ'', the polynomial $A'(\zeta' + \tau\zeta'')$ has exactly m roots $\tau_1^+(\zeta',\zeta''), \dots , \tau_m^+(\zeta',\zeta'')$ with positive imaginary part (and so, an equal number of roots with negative imaginary part).

For $n > 2$, each elliptic expression is properly elliptic. The same holds true when $n = 2$ if $A'(\zeta)$ has real coefficients and $r = 2m$.

Let Ω be a bounded domain in $\mathbb{R}^n$ with a smooth enough boundary Γ, and let $a_\alpha(t)$ be infinitely differentiable functions given on the closed domain $\bar{\Omega} = \Omega \cup \Gamma$. [To simplify the discussion, all results are formulated under maximal smoothness assumptions.] A differential expression

$$A(t,D) = \sum_{|\alpha|\leq r} a_\alpha(t) D^\alpha$$

is called *elliptic* (*properly elliptic*) in $\bar{\Omega}$ if, for each fixed $t_0 \in \bar{\Omega}$, the differential expression $\sum_{|\alpha|\leq r} a_\alpha(t_0)D^\alpha$ with constant coefficients is elliptic (properly elliptic).

The solvability theory of boundary value problems for elliptic differential equations is significantly more complicated than that presented in §20.1 for ordinary differential equations. There the proof of the Noetherian character of boundary value problems was based on the finite-dimensionality of the kernel of the differential operator and on the finite-dimensionality of the set of values of the additional (boundary) operators. Here these two "finite-dimensionalities" no longer exist: the set of solutions of a homogeneous elliptic equation is infinite dimensional, while the boundary operators act in infinite dimensional spaces of functions defined on the boundary of the domain.

Due to these circumstances, the corresponding boundary value problems are Noetherian only for certain classes of boundary conditions, which we shall describe now.

The restriction of a differential expression on $\bar{\Omega}$ to the boundary Γ,

$$B(s,D) = \sum_{|\beta|\leq m} b_\beta(t)D^\beta \Big|_{t=s\in\Gamma}$$

is called a *boundary* differential expression (the coefficients b_β are assumed to be infinitely differentiable also).

A system of m boundary differential expresssions

$$B_j(s,D) = \sum_{|\beta|\leq n_j} b_{j\beta}(s)D^\beta \quad (j = 1,\ldots,m)$$

is said to satisfy the *Shapiro-Lopatinskiĭ complementary condition* relative to the expression $A(t,D)$ if the following holds true. For each point s of the boundary Γ, denote by ν the unit outward normal to Γ at s, denote by ξ an arbitrary vector tangent to Γ at s, and denote by $\tau_i^+(s,\xi)$, $1 \leq i \leq m$, all the roots of the polynomial $A'(s,\xi + \tau\nu)$ which have positive imaginary part. Then the polynomials $B_j'(s,\xi + \tau\nu)$ in the variable τ must be linearly independent modulo the polynomial

$$\prod_{i=1}^{m} (\tau - \tau_i^+(s,\xi)) .$$

(In other words, there is no linear combination with nonzero coefficients of the polynomials B_j' which is divisible by the last polynomial).

A system of boundary expressions $B_j(s,D)$ $(j = 1,\ldots,m)$ is called *normal* if the orders n_j of these expressions do not exceed $2m-1$ and

$$\sum_{|\beta|=n_j} b_{j\beta}(s)\nu^\beta \neq 0$$

for any $s \in \Gamma$ and any vector ν normal to Γ at s.

Given any properly elliptic expression $A(t,D)$, the boundary operators

$$B_j(s,D) = \frac{\partial^j}{\partial \nu^j}$$

of the Dirichlet problem (where $\frac{\partial}{\partial \nu}$ is the derivative relative to the normal at point s) form a normal system satisfying the Shapiro--Lopatinskiĭ complementary condition.

A peculiarity of elliptic boundary value problems is that the theory "does not work" in the spaces $C(\bar{\Omega})$ and $L_1(\Omega)$. However, in spaces $L_p(\Omega)$ and Hölder spaces this theory is sufficiently developed.

For simplicity, the results that we discuss here are formulated in the space $L_2(\Omega)$ and those spaces related to it.

Consider the boundary value problem

$$A(t,D)x(t) = y(t) \quad \text{in} \quad \Omega \tag{20.12}$$

$$B_j(s,D)x(s) = \phi_j(s) \quad \text{on} \quad \Gamma \ (j = 1,\dots,m). \tag{20.13}$$

This problem generates an operator A_B, defined on all functions from $W_2^{2m}(\Omega) \subset L_2(\Omega)$, and given by the formula

$$A_B x = \{A(t,D)x \ ; \ B_1(s,D)x, \ \dots \ ,B_m(s,D)x\}$$

(here the derivatives are already understood as generalized derivatives). A_B takes its values in the product space

$$L_2(\Omega) \times W^{2m-n_1-1/2}(\Gamma) \times \dots \times W^{2m-n_m-1/2}(\Gamma). \tag{20.14}$$

Here $W_2^{k-1/2}(\Gamma)$ stands for the space of traces on Γ of all functions from $W_2^k(\Omega)$, with the natural norm.

The following statement is one of the basic facts in the theory of elliptic problems.

Let $A(t,D)$ *be a properly elliptic expression, and let* $B_j(s,D)$ *be a normal system of boundary operators* $(j = 1,\dots,m)$. *Then the operator* A_B *generated by the boundary value problem* (20.12), (20.13) *is Noetherian, as an operator from* $L_2(\Omega)$ *into the space* (20.14).

The kernel of the operator A_B *is finite dimensional and consists of infinitely differentiable functions.*

To describe the set of right-hand sides for which (20.12), (20.13) is solvable, we shall introduce the adjoint boundary value problem.

By definition, the *formal adjoint* (or *transpose*) of the differential expression $A(t,D)$ is the expression

$$A^{\dagger}(t,D)g = \sum_{|\alpha|\leq r} D^{\alpha}(\bar{a}_{\alpha}(t)g) .$$

Any normal system of boundary expressions $B_j(s,D)$ $(j = 1,\ldots,m)$ can be completed by a normal system of boundary expressions $S_j(s,D)$ of orders ν_j $(j = 1,\ldots,m)$ such that the orders $n_1,\ldots,n_m,\nu_1,\ldots,\nu_m$ of the system of expressions $\{B_1,\ldots,B_m,S_1,\ldots,S_m\}$ exhaust all the numbers from 0 to $2m-1$. Having the expressions B_j and S_j, one can construct, in a unique way, normal systems of boundary expressions $B_j^{\dagger}$ and $S_j^{\dagger}$ $(j = 1,\ldots,m)$ with the properties: $B_j^{\dagger}$ has order $2m-1-\nu_j$, $S_j^{\dagger}$ has order $2m-1-n_j$, and the following identity holds

$$\int_{\Omega}(Ax)\bar{g}dt + \sum_{j=1}^{m}\int_{\Gamma}(B_jx)\overline{S_j^{\dagger}g}ds = \sum_{j=1}^{m}\int_{\Gamma}(S_jx)\overline{B_j^{\dagger}g}ds + \int_{\Omega}x\overline{A^{\dagger}g}dt .$$

This identity is called *Green's formula*, and the system $B_j^{\dagger}(s,D)$ $(j = 1,\ldots m)$ is called the *system of adjoint* (or *transposed*) *boundary expressions* for the system $B_j(s,D)$.

Problem (20.12), (20.13) *is solvable for given right-hand sides* $(y(t);\{\phi_j(s)\})$ *belonging to the space* (20.14) *if and only if the orthogonality condition*

$$\int_{\Omega} y\bar{g}dt + \sum_{j=1}^{m}\int_{\Gamma}\phi_j\overline{S_j^{\dagger}g}ds = 0$$

is satisfied for all solutions of the adjoint homogeneous boundary value problem

$$A^{\dagger}(t,D)g = 0 \quad \text{in} \quad \Omega,$$

$$B_j^{\dagger}(s,D)g = 0 \quad \text{on} \quad \Gamma, \quad (j = 1,\ldots,m) .$$

We must mention here that, under certain circumstances, a boundary value problem is Noetherian even if the system of boundary operators is not normal. However, the Shapiro-Lopatiskiĭ complementary condition *must* be satisfied.

In a series of works over the last 10-12 years [Translator's note. The Russian version of the book appeared in 1971] the statements made above have been proved either by using the method of a priori estimates (see §§ 8,9; in the theory of elliptic equations, such estimates are called *coercive inequalities*), or by constructing regularizers (see

§§ 14,15). [Translator's note. In the theory of partial differential equations, regularizers are often called *parametrices*.] Sometimes, the two methods have been used together: the coercive inequalities - to prove the n-normality, and the construction of a right regularizer - to establish d-normality.

The boundary value problem (20.12), (20.13) *has zero index if and only if there exists* θ $(-\pi < \theta \leq \pi)$ *such that*

1. $\dfrac{A'(t,\zeta)}{|A'(t,\zeta)|} \neq e^{i\theta}$ *for all* $\zeta \neq 0$ *and* $t \in \bar{\Omega}$,

and

2. *the polynomials* $B_j'(s,\xi + \tau\nu)$ *in the variable* τ *are linearly independent modulo every polynomial*

$$\prod_{i=1}^{m} (\tau - \tau_i^+(s,\xi,\lambda)),$$

where $\arg \lambda = \theta$ *and* $\tau_i^+(s,\xi,\lambda)$ *are the roots of the polynomial* $A'(s,\xi + \tau\nu) - \lambda$ *in the upper half-plane.*

A solvability theory for systems of elliptic partial differential equations was also constructed. One of the most significant achievements of algebraic topology in the last years was to settle, theoretically, the problem of calculating the index of equations corresponding to boundary value problems for elliptic systems.

3. A BOUNDARY VALUE PROBLEM FOR AN EQUATION FROM VECTOR CALCULUS.

Consider in a domain Ω of three-dimensional space, the equations

$$\operatorname{rot} \vec{u} = \vec{v} , \quad \operatorname{div} \vec{u} = h , \tag{20.15}$$

where $\vec{v}$ is a given vector-valued function, h is a scalar function, and $\vec{u}$ is the unknown vector-valued function. We assume that $\vec{v} \in \vec{L}_2(\Omega)$ and $h \in L_2(\Omega)$. System (20.15) is overdetermined: the first equation cannot be solved for any right-hand side, because $\operatorname{div} \operatorname{rot} \vec{u} = 0$. Therefore, a necessary condition for the solvability of the first equation in (20.15) is $\operatorname{div} \vec{v} = 0$.

The closure in $\vec{L}_2(\Omega)$ of the space of all solenoidal vector-valued functions (i.e., functions $\vec{v}$ satisfying $\operatorname{div} \vec{v} = 0$) is a subspace which we denote by $\vec{S}$. Therefore, the solvability of equations (20.15) may be discussed only in the subspace $\vec{S} \times L_2(\Omega)$ of $\vec{L}_2(\Omega) \times L_2(\Omega)$.

The orthogonal complement to $\vec{S}$ in $\vec{L}_2(\Omega)$ is the subspace of gradients of scalar functions which vanish on the boundary Γ of the domain Ω. Due to this, one may employ the device described at the end of § 17, and consider the system

$$\left.\begin{aligned} \operatorname{rot} \vec{u} + \operatorname{grad} p &= \vec{v}, \\ \operatorname{div} \vec{u} &= h, \end{aligned}\right\} \tag{20.16}$$

with the condition $p_{|\Gamma} = 0$ and also some other boundary condition

$$B\vec{u} = \phi \quad \text{on} \quad \Gamma. \tag{20.17}$$

To the various solvability properties of system (20.16), (20.17) correspond similar solvability properties of the boundary value problem (20.15), (20.17) in the subspace introduced above. System (20.16) happens to be elliptic, and the corresponding boundary value problem is Noetherian if and only if the Shapiro-Lopatinskiĭ complementary condition is satisfied. Consequently, we obtain a necessary and sufficient condition for problem (20.15), (20.17) to be Noetherian in a subspace of the corresponding space (which we do not specify).

This condition is as follows:

Let s *be an arbitrary point of the boundary* Γ, *and let* $(\sigma_1, \sigma_2, \nu)$ *be a system of local coordinates in a neighborhood of* s, *where the axes* σ_1, σ_2 *lie in the tangent plane to* Γ, *and* ν *lies along the normal to* Γ. *If the boundary condition* (20.17) *has, in these coordinates, the form*

$$\sum_{j=1}^{3} B_j\left(s, i\frac{\partial}{\partial\sigma_1}, i\frac{\partial}{\partial\sigma_2}, i\frac{\partial}{\partial\nu}\right)u_i = 0$$

(*where* u_i *are the components of* u *in coordinates* σ_1, σ_2, ν, *then*

$$\sum_{j=1}^{3} B_j(s, \xi_1, \xi_2, \xi_3) \neq 0$$

for all real ξ_1, ξ_2, $\xi_1^2 + \xi_2^2 \neq 0$, *and* $\xi_3 = i(\xi_1^2 + \xi_2^2)^{1/2}$.

4. THE EXTERIOR DIRICHLET PROBLEM FOR THE HELMHOLTZ EQUATION.

Let D be a domain in the three-dimensional space $\mathbb{R}^3$ bounded by a smooth enough closed surface Γ, and let $\Omega = \mathbb{R}^3 \setminus \bar{D}$. One must find a function u satisfying the Helmholtz equation

$$\Delta u(t) + k^2 u(t) = 0 \tag{20.18}$$

in Ω, the boundary condition

$$u(s)_{|\Gamma} = h(s) , \tag{20.19}$$

and the radiation condition at infinity

$$u = O(\rho^{-1}), \quad \frac{\partial u}{\partial \rho} - iku = o(\rho^{-1}) \quad \text{for} \quad \rho = |t| \longrightarrow \infty . \tag{20.20}$$

The solution of this problem is sought as the sum of the simple and double layer potentials, which leads to a undetermined equation in $L_2(\Omega)$ relative to their densities (see [33]).

APPENDIX

BASIC RESULTS FROM FUNCTIONAL ANALYSIS USED IN THE TEXT

1. THE OPEN MAPPING PRINCIPLE. *A continuous linear map of one Banach space into another takes any open set into an open set.*

As corollaries of this principle we have:

1) If a linear operator A from a Banach space E into a Banach space F is one-to-one, onto, and continuous, then the inverse operator A^{-1} is continuous.

2) Let a space E be endowed with two norms $\|x\|_1$ and $\|x\|_2$ which satisfy $\|x\|_1 \leqq C\|x\|_2$, $\forall\, x \in E$. If E is a complete (Banach) space in both norms, then the inverse equality $\|x\|_2 \leqq C_1\|x\|_1$ holds true for all $x \in E$, i.e., the two norms are equivalent.

Another consequence is actually proved in the text.

3) If a closed operator A is defined on the entire Banach space E, then it is bounded (see § 2).

2. HAHN-BANACH THEOREM. *Any linear bounded functional given on a linear submanifold* L *of a Banach space* E *can be extended, preserving its norm, to a linear bounded functional on the entire space* E.

The following statements are particular cases of this theorem.

1) Let R be a subspace of a Banach space E, and let $x_0 \notin R$. Then there is a bounded linear functional $f(x)$, equal to zero on R and such that $f(x_0) = 1$.

2) Given any element $x_0 \in E$, there is a linear functional $f(x)$ such that $\|f\| = 1$ and $f(x_0) = \|x_0\|$.

3) Mazur's theorem. *Let* Q *be a closed, convex subset of a Banach space* E, *and let* $x_0 \notin Q$. *Then there is a bounded linear functional,* $f(x)$, *on* E, *such that* f *separates* x_0 *and* Q, *i.e.,* $f(x_0) > \sup_{x\in Q} f(x)$.

If, additionally, Q *is balanced, i.e., if* $x \in Q$ *implies* $-x \in Q$, *then one can choose* $f(x)$ *to satisfy* $f(x_0) > \sup_{x\in Q} |f(x)|$.

3. UNIFORM BOUNDEDNESS PRINCIPLE. *Let* E *be a Banach space, and let* $\{\Phi_\alpha(x)\}$ *be a family of continuous functionals, each of them satisfying* (*the convexity conditions*)

$$1^\circ.\ \ \Phi_\alpha(x) \geqq 0\ ;\quad 2^\circ.\ \ \Phi_\alpha(\lambda x) = |\lambda|\Phi_\alpha(x)\ ;$$

$$3^\circ.\ \ \Phi_\alpha(x+y) \leqq \Phi_\alpha(x) + \Phi_\alpha(y)\ .$$

If $\sup\limits_\alpha \Phi_\alpha(x) \leqq k(x) < \infty$ *for any fixed* $x \in E$, *then there is a constant* C *such that*

$$\Phi_\alpha(x) \leqq C\|x\|\ ,\quad \forall\ x \in E.$$

On the basis of the uniform boundedness principle one can prove the following statements.

1) If a sequence of bounded linear functionals f_n converges weakly to a functional f, i.e., $f_n(x) \longrightarrow f(x)$ for all $x \in E$, then the limit functional f is linear and bounded.

2) Banach-Steinhaus Theorem. *A sequence of bounded linear functionals on a Banach space* E *is weakly convergent if and only if it converges on some dense subset of* E *and the sequence of norms of these functionals is bounded.*

4. SUBSPACES. If L is a linear manifold in a linear space E, then there always exists an algebraic direct complement to L, i.e., a linear manifold M such that $E = L + M$ and $L \cap M = \{\theta\}$. In other words, each element $x \in E$ can be expressed uniquely as $x = u + v$, with $u \in L$, $v \in M$.

However, a subspace L of a Banach space does not necessarily have a closed direct complement. [Translator's note. We tried to adhere to the convention: a *subspace* is a *closed* linear manifold.]

If $E = L \oplus M$, where L, M are subspaces of E, then the norm

$$\|x\|_1 = \|u\| + \|v\| \qquad (x = u + v,\ \ u \in L,\ \ v \in M)$$

is equivalent to the initial norm $\|x\|$ on space E. (This can be proved as a consequence of the open mapping principle.)

Riesz's Theorem. *A subspace is locally compact if and only if it is finite dimensional.*

Any finite dimensional subspace has a closed direct complement. Moreover, if L is a finite dimensional subspace and R is a subspace such that $L \cap R = \{\theta\}$, then there is a closed direct complement M to L, such that $M \supset R$.

If L is a finite dimensional subspace and $x_0 \notin L$, then there is an element $y \in L$ which is the nearest to x_0, i.e., such that $\| x_0 - y\| = \rho(x_0, L) = \inf_{z \in L} \| x_0 - z\|$.

If L is a finite dimensional subspace and R is any subspace, then $L + R$ is a subspace.

5. QUOTIENT SPACE. If N is a subspace of a normed space E, then the quotient (or factor) space E/N, i.e., the space of cosets $X = x + N$ with the natural operations, is a normed space with the norm

$$\| X\| = \inf_{z \in N} \| x + z\| .$$

If E is a Banach space, then E/N is also Banach (see the proof in § 1). If the subspace N has a closed direct complement M, then the quotient space E/N is isomorphic to M. If the space E/N is finite dimensional, then N has a closed direct complement of dimension $\dim E/N$ (see § 8).

6. DUAL SPACE. The set of all bounded linear functionals on a Banach space E becomes, after one introduces the natural algebraic operations, a linear space E^*. This space is also normed : $\| f\| = \sup_{\| x\| = 1} |f(x)|$. Relative to this norm, E^* is a Banach space, called the *dual* (or *conjugate*) space of E.

In the dual space E^*, one can additionally introduce the weak topology. A sequence of linear functionals converges in this topology if it converges weakly (i.e., $f_n(x) \longrightarrow f(x)$, $\forall x \in E$). If the linear manifold $\Gamma \subset E^*$ is total, i.e., if for any $x \in E$ there is a functional $f \in \Gamma$ such that $f(x) \neq 0$, then the weak closure of Γ is E^*.

Each element $x \in E$ defines, in a natural way, a linear functional F_x on E^*, by the formula $F_x(f) = f(x)$ $(f \in E^*)$. In other words, the space E can be linearly and isometrically embedded into the space E^{**}. A space E is called *reflexive* if its image under this embedding equals E^{**}.

If the space E is not reflexive, then a bounded linear functional $F(f)$ on E^* has the form $F_x(f)$ with $x \in E$ if and only if $F(f)$ is continuous on E^* in the weak topology.

In the dual space E^* of a reflexive space E, the weak and strong closures of a linear manifold coincide.

7. ORTHOGONAL COMPLEMENTS. If N is a linear manifold in the

Banach space E, then its *orthogonal complement* $N^{\perp}$ is, by definition, the set of all bounded linear functionals $f \in E^*$ such that $f(x) = 0$ for all $x \in N$. The orthogonal complement $N^{\perp}$ is a subspace of the dual space E^*.

If S^* is a linear manifold in E^*, then one can consider two orthogonal complements for it: the set ${}^{\perp}S^*$ of all elements $x \in E$ such that $f(x) = 0$ for all $f \in S^*$, and the set $S^{*\perp}$ of all elements $F \in E^{**}$ such that $F(f) = 0$ for all $f \in S^*$. There is a natural embedding ${}^{\perp}S^* \subset S^{*\perp}$. If ${}^{\perp}S^*$ is finite dimensional, then ${}^{\perp}S^* = S^{*\perp}$. If E is reflexive, the same holds, i.e., ${}^{\perp}S^* = S^{*\perp}$.

The equality ${}^{\perp}(N^{\perp}) = \bar{N}$ is also valid (see §3 for a proof).

Given a linear manifold $S^* \subset E^*$, the linear manifold $({}^{\perp}S^*)^{\perp}$ does does not necessarily coincide with $\overline{S^*}$. If S^* or ${}^{\perp}S^*$ is finite dimensional, then $({}^{\perp}S^*)^{\perp} = \overline{S^*}$. In general, $({}^{\perp}S^*)^{\perp}$ equals the weak closure of the linear manifold S^*.

LITERATURE CITED

BASIC

Monographs and Textbooks

[1] S. Banach, Théorie des Opérations Linéaires, Monografje Matematyczne, Warsaw, 1932; Ukrainian transl., A Course in Functional Analysis, Kiev, 1948.

[2] N. Bourbaki, Éléments de Mathématique, Livre V, Espaces Vectoriels Topologiques, Act. Sci. et Ind., 1189, 1229, Hermann et Cie., Paris, 1953, 1955; Russian transl., IL, Moscow, 1959.

[3] N. Dunford and J. Schwartz, Linear Operators. I: General Theory, Pure and Appl. Math., vol. 7, Interscience, New York and London, 1958; Russian transl., IL, Moscow, 1962.

[4] I. C. Gohberg and M. G. Krein, Introduction to the Theory of Linear Nonselfadjoint Operators, Nauka, Moscow, 1965; English transl., Transl. Math. Monographs, vol. 18, Amer. Math. Soc., Providence, R.I., 1969.

[5] S. Goldberg, Unbounded Linear Operators. Theory and Applications, McGraw Hill, London, Toronto, New York, 1966.

[6] F. Hausdorff, Mengenlehre, 3. Aufl., Dover Publications, New York, 1948; Russian transl., ONTI, 1936.

[7] E. Hille and R. S. Phillips, Functional Analysis and Semigroups, rev. ed., Amer. Math. Soc. Colloq. Publ., vol. 31, Amer. Math. Soc., Providence, R.I., 1957; Russian transl., IL, Moscow, 1962.

[8] L. Hörmander, An Introduction to Complex Analysis in Several Variables, Van Nostrand, Princeton, N.J., Toronto, New York, London, 1966; Russian transl., Mir, Moscow, 1968.

[9] L. V. Kantorovich and G. P. Akilov, Functional Analysis in Normed Spaces, Fizmatgiz, Moscow, 1959; English transl., Pergamon Press, Oxford, 1964.

[10] T. Kato, Perturbation Theory for Linear Operators, Die Grundlehren der Matem. Wissenschaften, Band 132, Springer-Verlag, Berlin, Heidelberg, New York, 1966; Russian transl., Mir, Moscow, 1972.

[11] L. A. Liusternik and V. I. Sobolev, Elements of Functional Analysis, Nauka, Moscow, 1965; English transl., Ungar, New York, 1961.

[12] S. G. Mikhlin, Multidimensional Singular Integrals and Integral Equations, Fizmatgiz, Moscow, 1962; English transl., Pergamon Press, Oxford, 1965.

[13] D. Przeworska-Rolewicz and S Rolewicz, Equations in Linear Spaces, PWN, Warsaw, 1968.

[14] F. Riesz and B. Sz.-Nagy, Lecons sur l'Analyse Fonctionnelles, Akadémiai Kiadó, Budapest, 1972; Russian transl., Mir, Moscow, 1979.

[15] K. Yosida, Functional Analysis, Die Grundlehren der Matem. Wissenschaften, Band 123, Springer-Verlag, Berlin, Göttingen, Heidelberg, 1965; Russian transl., Mir, Moscow, 1967.

Articles

[16] F. V. Atkinson, Normal solvability of linear equations in normed spaces, Mat. Sb. 28 (1951), 3-13. (Russian)

[17] F. V. Atkinson, On relatively regular operators, Acta Sci. Math., 15 , 1 (1953), 38-56.

[18] F. E. Brower, Functional Analysis and partial differential equations, Math. Ann. 138 (1959), 55-79.

[19] J. Dieudonne, Sur les homomorphismes d'espaces normes, Bull. Sci. Mat. (2) 67 (1943), 72-84.

[20] T. W. Gamelin, Decomposition theorems for Fredholm operators, Pacific J. Math. 15 , 1 (1965), 97-106.

[21] I. C. Gohberg, On linear equations in Hilbert space, Dokl. Akad. Nauk SSSR 76, No. 1 (1951), 9-12. (Russian)

[22] ——, On linear equations in normed spaces, Dokl. Akad. Nauk SSSR 76 , No. 4 (1951), 477-480. (Russian)

[23] ——, On the index of an unbounded operator, Mat. Sb. 33 (1953), 193-198. (Russian)

[24] ——, On zeroes and null-elements of unbounded operators, Dokl. Akad. Nauk SSSR 101, No. 1 (1955), 9-12. (Russian)

[25] ——, Two remarks on the index of a linear bounded operator, Uč. Zap. Bel'ckogo Ped. Inst. 1 (1958), 13-18. (Russian)

[26] I. C. Gohberg and M. G. Krein, The basic propositions on defect numbers, root numbers, and indices of linear operators, Uspehi Mat. Nauk 12 (1957), 43-118 (Russian); English transl., Amer. Math. Soc. Transl. (2) 13 (1960), 185-264.

[27] S. Goldberg, Linear operators and their conjugates, Pacific J. Math. 9, 1 (1959), 69-79.

[28] ——, Closed linear operators and associated continuous linear operators, Pacific J. Math. 12, 1 (1962), 808-811.

[29] M. A. Gol'dman, On the stability of the normal solvability property for linear operators, Dokl. Akad. Nauk SSSR 100 , No. 2, (1955), 201-204. (Russian)

[30] J. T. Joichi, On operators with closed range, Proc. Amer. Math. Soc., 11 (1960), 80-83.

[31] T. Kato, Perturbation theory for nullity, deficiency and other quantities of linear operators, J. Analyse Math. 6 (1958), 361-422.

[32] M. A. Krasnosel'skii and M.G. Krein, Stability of the index of an unbounded operator, Mat. Sb. 30 (1952), 219- 224. (Russian)

[33] M. G. Krein, On one undetermined equation in Hilbert space and its application to potential theory, Uspehi Mat. Nauk 9, no. 3 (1954), 149-153. (Russian)

[34] S. G. Mikhlin, On the solvability of linear equations in Hilbert space, Dokl. Akad. Nauk SSSR 57, no.1 (1947), 11-12. (Russian)

[35] ——, Two theorems on regularizers, Dokl. Akad. Nauk SSSR 125, no. 2 (1959), 278-280. (Russian)

[36] B. Sz.-Nagy, Perturbations des transformations lineaires fermées, Acta. Sci. Math. 14, no. 2 (1951), 125-137.

[37] ——, On the stability of the index of unbounded linear transformations, Acta Sci. Math. 3, no. 1-2 (1952), 49-52.

[38] S. M. Nikol'skiĭ, Linear equations in linear normed spaces, Izv. Akad. Nauk SSSR Ser. Mat. 7, no. 3 (1943), 147-166. (Russian)

[39] J. Petree, Another approach to elliptic boundary value problems, Comm. Pure Appl. Math. 14 (1961), 711-731.

[40] S. Prössdorf, Operators which admit unbounded regularization, Vestnik LGU Ser. Mat. Mech. Astr. 13, no. 3 (1965), 59-67. (Russian)

[41] D. Przeworska-Rolewicz and S. Rolewicz, On operators with finite d-characteristic, Studia Math. 24 (1964), 257-270.

[42] ——, On operators preserving a conjugate space, Studia Math. 25 (1965), 251-255.

[43] ——, On quasi-Fredholm ideals, Studia. Math. 26 (1965), 67-71.

[44] ——, On d- and d_E-characteristics of linear operators, Ann. Polon. Math. 19 (1967), 117-121.

[45] M. Schechter, Basic theory of Fredholm operators, Ann. Scuola Norm. Super. Pisa, Sci. Fis. e Math., ser. 3, 21 (1967), 261-280.

[46] A. E. Taylor and Ch. J. A. Halbury, General theorems about a bounded linear operator and its conjugate, J. Reine Angew. Math. 198 (1957), 93-111.

[47] I. S. Tetievskaja, On a undetermined equation in Banach space, Mat. Issl. IV, no. 1 (1969), 171-173. (Russian)

[48] B. Yood, Properties of linear transformations preserved under addition of a completely continuous transformation, Duke Math. J. 18 (1951), 599-612.

Supplementary literature

[49] S. Agmon, A. Douglis, and N. Nirenberg, Estimates near the boundary for solutions of elliptic partial differential equations satisfying general boundary conditions, I, Comm. Pure Appl. Math. 12 (1959), 623-727; ibid., II, Comm. Pure Appl. Math. 17 (1964), 35-92. Russian transl., IL, Moscow, 1962.

[50] G. I. Eskin and M. I. Vishik, Elliptic equations in convolution in a bounded domain and their applications, Uspehi Mat. Nauk 32, no. 1 (1967), 15-76 (Russian); English transl., Russian Math. Surveys 12 (1967), 13-75.

[51] I. C. Gohberg and M. G. Krein, Systems of integral equations on a half-line with kernels depending on the difference of arguments, Uspehi Mat. Nauk 13, no. 2 (1958), 3-72 (Russian); English transl., Amer. Math. Soc. Transl. (2), 14 (1960), 217-287.

[52] M. G. Krein, Integral equations on a half-line with kernel depending on the difference of argumaents, Uspehi. Mat. Nauk 13, no. 5 (1958), 3-120 (Russian); English transl., Amer. Math. Soc. Transl. (2), 4 (1958), 127-131.

[53] N. I. Muskhelishvili, Singular Integral Equations, Nauka, Moscow, 1968; English transl., Noordhoff, Groningen, 1953.

[54] M. A. Naimark, Linear Differential Operators, Nauka, 1969; English transl., Ungar, New York, 1967-68.

[55] V. A. Solonnikov, General boundary value problems for Douglis--Nirenberg elliptic systems. I, Izv. Akad. Nauk SSSR Ser. Mat. 28 (1964), 665-706; English transl., Amer. Math. Soc. Transl. (2) 56 (1966), 193-232, and II, Trudy Mat. Inst. Akad. Nauk SSSR im. Steklova, 92 (1966), 233-297; English transl., Proc. Steklov Inst. Math., 92 (1966), 269-339.

[56] L. R. Volevič, Solvability of boundary value problems for general elliptic systems, Mat. Sb. 68 (1965), 373-416; English transl., Amer. Math. Soc. Transl. (2) 67 (1968), 182-225.

[57] P. P. Zabreyko et al., Integral Equations, SMB, Nauka, Moscow, 1968; English transl., Noordhooff, Leyden, 1975.

Basic Operator Theory

Israel Gohberg, *Tel-Aviv University, Israel*
Seymour Golberg, *University of Maryland, College Park*

Basic Operator Theory provides an introduction to functional analysis with an emphasis on the theory of linear operators and its application to differential and integral equations, approximation theory, and numerical analysis. It is designed primarily for use as a textbook by senior undergraduate and graduate students, and is presented as a natural continuation of linear algebra. **Basic Operator Theory** provides a firm foundation in operator theory, an essential part of mathematical training for students of mathematics, engineering, and other technical sciences.

Contents (General Chapter Headings)

Appendices

NOTES

NOTES

NOTES

NOTES

NOTES

NOTES

NOTES

NOTES

NOTES

NOTES